The Energy Trap

The Energy Trap

Why the renewable energy transition can't work – and what can

Richard Lyon

FORUM

FORUM

First published in Great Britain by Forum,
an imprint of Swift Press 2026

1 3 5 7 9 8 6 4 2

Printed and bound in Great Britain by CPI Group (UK) Ltd, Croydon CR0 4YY

A CIP catalogue record for this book is available from the British Library

ISBN: 9781800757356
eISBN: 9781800757363

MIX
Paper | Supporting responsible forestry
FSC® C013604
FSC www.fsc.org

To my sons, Alexander, Benjamin and Christopher,
to their children, and to their children's children.

Contents

Prologue p.1
Introduction p.5

PART I
The Physics of Energy
p.9

PART II
The Industrial Metabolism
p.25

PART III
The Depleting Inheritance
p.33

PART IV
The Renewables Paradox
p.51

PART V
The Escape Hatches
p.69

PART VI
Energy and Your Money
p.79

PART VII
Forging a New Realism
p.89

Epilogue p.111
What's Next? p.113
Glossary p.115
Notes p.121
Further Reading p.127

Prologue

Fortunately, there aren't many threats in our lives that can kill large numbers of people, and most of them sound like Hollywood disaster movies: tsunamis, escaped viruses, asteroids, volcanoes.

So it's surprising that one of the biggest threats we now face comes from an area of life we thought we'd thoroughly mastered – one that, for over a hundred years, has delivered rising prosperity, increasing lifespans and the elimination of famine from the developed world: our energy system. That energy system, on which billions of lives literally depend, is now being dismantled – with confidence, with public money and with almost no scrutiny.

Think that's an exaggeration? Here's an example. Roughly half the world's population – four billion people – depend for life on food grown with fossil fuel-based fertilisers. There aren't any significant stockpiles of either food or fertiliser. So if we were to stop oil and gas production today, as some climate activists demand, then within a decade, hundreds of millions of people would starve to death. Yet that is precisely the direction in which the energy policies of advanced industrial countries like the United Kingdom now point.

The effect, for those with eyes to see, is rather like looking out of the window of your aircraft and noticing someone methodically sawing through the wing. You want to turn to the people around you and say: 'Can you see what's happening? We have to stop this.'

And that is why I wrote this book.

I wrote it because too many people can't yet see the saw. Worse, quite a few have been persuaded that sawing through the wing is a good idea – that the energy system must be sacrificed to save the climate.

There are many excellent books about energy systems. I've noted a few in the 'Further Reading' section. They run to hundreds of pages, they're written for specialists and they sit unread on shelves while the policy rolls on. What's missing is something shorter and sharper – a friendly guide that gives you the minimum set of facts you need, with references so you can check the claims yourself, written by someone who understands the subject and wants to explain it clearly. That's what this book tries to be.

And why am I the person to write it? Let me tell you a little about my journey. My father never quite understood my career, and I can't blame him – I didn't plan it. But it produced an unusual combination of education, knowledge and experience that turns out to be what you need to understand this subject. My first degree was in Electrical Engineering. After serving in the Royal Air Force, I took another in Petroleum Engineering and, armed with that, joined a large oil and gas company and spent two decades managing projects around the world.

Large oil companies have a habit of throwing unexpected things at you. One day I was writing briefings for senior people at Headquarters. 'Now see here, Lyon. We keep getting these questions at shareholder meetings about something called "Peak Oil". Give me the bullets. I'll fire the gun. Write me a memo.' I started writing the memo. The memo turned into a master's degree in Energy Economics and a good understanding of the relationship between the energy system and the financial system (spoiler: we get it backwards).

That's the journey. Electrical engineering gave me maths, physics, chemistry and electrical systems. Petroleum engineering showed me what it takes to keep the energy flowing. Energy economics revealed the central error: the assumption that low-quality energy – diffuse, intermittent, hard to concentrate – can simply replace high-quality energy because the price is right.

Draw a Venn diagram of those three, and right in the centre is one of the biggest threats of our lifetimes: the policy of dismantling our energy system before anything remotely adequate exists to replace it. I read the early proposals for a 'renewable energy transition' as they were beginning to circulate in policy circles, and couldn't believe my eyes. Then I looked at who was writing them: people with no training in physics, no engineering, no understanding of power systems, no experience of producing energy and no grasp of the economics of finite resource extraction. A policy consensus formed by people who understood none of the disciplines needed to evaluate what they were proposing. They didn't know what they didn't know – and they didn't know that they didn't know it.

But this book isn't for them. It's for you – the person on whom this policy is being inflicted. You're perfectly capable of understanding what's happening if someone takes the trouble to explain it clearly.

You may not be fully aware yet of your dependence on the energy system and the significance of what is being done to you. That's understandable – until now, it's worked so well that you've never had to think about it. But that is also what makes it a trap – you cannot see it until you try to leave it, and by then the options have narrowed. This book is my attempt to show you the trap.

I've taken a lot of care to make it short. You can read it in three cups of tea and half a packet of biscuits. But when you're done, you'll understand more about energy than most of the people making energy policy. And you'll be able to say, 'I can see what's happening; we have to stop this', to the people with the saw.

Thank you for reading. I'll see you on the other side.

Richard Lyon
Edinburgh

Introduction

This is a little book about a big subject. It is written for the general reader who suspects that the public conversation about energy has drifted some distance from physical reality and would like a concise, honest account of where things stand.

All governments have an energy policy. But energy policy is not like other sorts of policy. Energy is the master resource, the precondition for every other activity that an industrial civilisation undertakes. The food on your table, the roof over your head, the hospital that treats you, the phone in your hand, even (as we shall see) the £10 note in your wallet – all these are, at bottom, transformed energy, or a claim on a future quantity of it. When we get energy wrong, we get everything wrong.

Governments have always understood this, at least implicitly. Every major energy transition in history – wood to coal, coal to oil – was driven by the discovery of a denser, more concentrated source, and the discovery of nuclear fission opened the same logical next step. The one now proposed is different. It asks us to reverse the direction. But it is a trap – no human society in history has ever achieved this.

The mainstream debate about how to attempt it is strikingly poor. It is dominated by competing narratives – techno-optimist, degrowth, green-growth, fossil-loyalist – each of which selects the facts that flatter its conclusions and ignores the rest. The result is that well-meaning people, including many in government, are making consequential decisions on the basis of claims that are not consistent with basic physics.

This book attempts something different. Rather than asking, 'What energy narratives do I prefer?' it starts from the physics and asks, 'What energy narratives are physically possible?' If a policy or a technology cannot satisfy the non-negotiable laws that govern how energy is captured, converted, stored and dissipated, it does not matter how elegant the spreadsheet or how passionate the advocacy: it will fail. Physical reality does not care about your feelings, your votes or your investment portfolio.

The argument is organised in seven parts. Part I asks the simplest question: what are the physical rules that every energy source must obey? The answer turns out to explain most of what is wrong with current energy policy. Part II asks what all this energy is actually *for*, and why the four industrial foundations of modern life – steel, cement, ammonia and plastics – cannot simply be electrified away. Part III asks what is happening to the fossil fuels themselves – the depleting inheritance that no energy policy acknowledges. Part IV turns to the proposed solution and shows why it won't work: the storage problem, the mineral hunger and the replacement costs that its advocates prefer not to discuss. Part V tests the three escape hatches most commonly offered – efficiency, hydrogen and green growth – and finds that none of them works. Part VI follows the money, asking what happens to a financial system built on the assumption of perpetual growth when the energy to fuel that growth is forced to contract. Part VII is about what comes next, and why nuclear power is the only technology consistent with the physics; our available choices about what lies ahead of us; and a seven-point blueprint for an energy policy grounded in reality rather than wishful thinking.

Each section is deliberately brief. The aim is not encyclopaedic coverage but a clear survey of the problem. A glossary, references and a list of further reading are provided for those who wish to dig deeper.

A word on tone. This book is not neutral, and it does not pretend to be. It takes the position that honesty about physical constraints is not pessimism but the necessary starting point for any serious response. If that strikes you as gloomy, consider the alternative: continuing to promise the public outcomes that the laws of physics will not deliver, and discovering the gap between promise and reality only when it is too late.

Like it or not, an energy transition is coming. The

question is whether it is one that is forced upon us by depletion and wishful thinking, or one that we have chosen and shaped. This book is written in the hope that it is not yet too late for the latter.

Part I

The Physics of Energy

'Ye cannae change the laws of physics, Captain!'
— Montgomery Scott, chief engineer, USS *Enterprise*

OK. I get it. Physics. Yuck.

Here's the thing. Energy is not particularly difficult to understand, at least at the level required to understand energy policy. But nor does it lend itself particularly well to guesswork.

To understand energy with proper intuition, you need to understand the physics of energy. Luckily, there are only three concepts and one critical distinction that you need to learn to understand 90 per cent of energy policy: energy gradient, energy density, areal power density and the difference between a *fuel* and an *energy carrier*. This is perhaps the hardest part of the book. But the payoff is enormous. Once you understand these concepts, most energy policy becomes transparent – including the parts that its advocates would prefer you not to examine too closely.

Ready?

ENERGY GRADIENT

The first concept is the most fundamental, and the most revealing. It explains why not all energy is equal – and why the sheer quantity of energy available tells you almost nothing about how much useful work you can extract from it.

It works like this. To do useful work, energy must flow from a region of high concentration to a region of low concentration. A machine can then be placed in the path of that flow to capture it. Let's call the difference in concentration the *energy gradient*.* The steeper the gradient, the more work you can extract. A steep gradient means useful energy; a shallow gradient means unusable energy.

Let's make this concrete. You can boil an egg in a pan but not in a swimming pool, even though there is far more energy in the pool. This is because the energy in the pan is concentrated in a couple of litres at 100°C, but in the pool it is spread through 2,500 cubic metres of water at 20°C. The egg doesn't 'see' the total energy in the pool. It sees the temperature gradient at its shell – and in the pool, it is too shallow.†

But could you steepen the gradient by concentrating the pool's energy into something useful? In principle, yes. A device called a heat pump – essentially a fridge in reverse – can move heat from a cooler source to a warmer one. Heat pumps work, and they work efficiently. But the heat pump must be manufactured, installed and powered – and all of that requires energy. Where does *that* energy come from? If it comes from a concentrated source like gas or

* In thermodynamics, the formal term for the work-available portion of energy is exergy. Our 'energy gradient' is an informal way of expressing the same idea: the greater the difference between source and surroundings, the more work you can extract.

† For the right task, of course, the pool's energy is perfectly usable. If you wanted to melt 10,000 ice cubes instead of boiling an egg, you would use the pool, not the pan – because now the gradient runs the other way.

nuclear, then the concentrated source is doing the work, not the pool. And if it comes from the pool, then you need another device to concentrate *that* – which needs energy of its own. Each step in the chain costs energy, and when the original source is diffuse, the cost of concentrating it consumes most of what you end up collecting. You haven't solved the problem. You've moved it.

A gas flame at 1,500°C in a 15°C room creates a steep gradient – an enormous concentration difference that can drive turbines, smelt steel and power an industrial economy. A swimming pool in a back garden produces a shallow one – there is real energy, but no concentration difference to exploit. Quantity matters, of course – a single match won't power a city. But quantity without concentration is just a lukewarm swimming pool. Usable energy is not a thing you have. It is a potential between two energy states – and it is the gradient between them that determines whether it can do useful work.

ENERGY DENSITY

So, gradient tells us whether an energy source can do useful work. But there is a second question that matters just as much: how much of that work can you concentrate into a small enough package to run a civilisation on? A society that can carry its energy with it – in a ship's hold, a fuel tank, a pipeline – can project power, move goods and sustain complexity. A society that cannot is stuck. The concept that governs this is energy density.

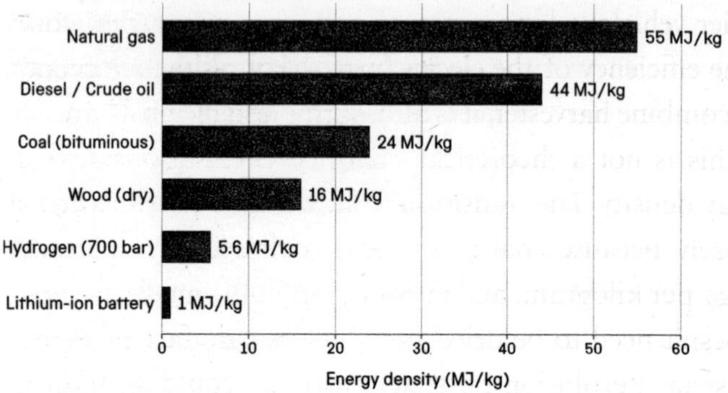

Fig.1 Energy density of common sources and storage media in megajoules per kilogram. The ratio between diesel and a lithium-ion battery – roughly 44:1 – is the single most important number in the energy debate.

Energy density measures how much energy is packed into a given mass or volume of fuel. It is the reason you can drive from London to Edinburgh on fifty litres of diesel, but you need a battery weighing roughly half a tonne to do the same thing in an electric car. The diesel contains roughly forty-four megajoules per kilogram. The best lithium-ion battery on the market manages about one megajoule per kilogram (see Figure 1). That is a ratio of 44:1. Engineering has improved batteries, but the gap remains enormous – because it is not an engineering problem. It is a chemistry problem. The carbon-hydrogen bonds in fossil fuels release a great deal of energy when broken. Shuttling lithium ions between electrodes releases much less. The periodic table is not subject to software updates. For a car, this is an inconvenience – a heavier vehicle, a longer stop to recharge, only partly offset by the efficiency of the electric motor. For a ship, an aircraft or a combine harvester, it is an insurmountable problem.

This is not a theoretical curiosity. Civilisations run on energy density. The transition from wood to coal succeeded precisely because coal is denser – with roughly twice the energy per kilogram, and far more per cubic metre, because it doesn't need to be dried first. That single fact made the Industrial Revolution possible. Factories could be built in cities rather than next to forests. Railways became possible.

The subsequent transition from coal to oil was another leap up the quality ladder. Oil packs about forty-four megajoules per kilogram versus coal's twenty-four – another doubling up the ladder. Oil is also a liquid, which means it can be pumped, piped and poured into a tank in minutes. That combination of high density and easy handling is what made aviation possible, what made the modern military possible and what made the globalised supply chain possible. None of these things can run on coal, and none of them can run on batteries – not because the batteries are bad, but because the physics won't allow it.

Here is the pattern. Every successful energy transition in human history has moved *up* the energy density ladder: wood to coal, coal to oil. The discovery of nuclear fission made the next step on that ladder available, but the world has not taken it at scale. Each step concentrated more energy into less mass and less space, enabling new capabilities that were physically impossible at the previous level.

Why is the density gap between hydrocarbons and renewables so large – and why can't engineering close it? Three factors, each compounding the others.

The first is the physics of the energy itself. Not all energy is stored in the same way. Nuclear fuel holds energy in the bonds between particles in an atomic nucleus – the strongest force in nature. Fossil fuels hold it in chemical bonds between carbon and hydrogen atoms – the next-strongest. Both lock enormous quantities of energy into tiny quantities of matter, available on demand. Solar radiation and wind are different in kind. They are flows, not stores – energy passing through, not energy locked in. You cannot pick up sunlight and put it in a tank. To use it, you must scavenge it, convert it and store it in something yourself.

The second is what we might call the compression of time and space. The diesel in your fuel tank is stored sunlight. Oil is the product of millions of years of photosynthesis, captured across millions of square miles of ancient forest and ocean, then compressed, cooked and concentrated for us by colossal geological forces over hundreds of millions of years – all for free. When you burn a litre of diesel, you are releasing a quantity of ancient sunlight that no solar panel could collect in a human lifetime. Wind turbines and solar farms scavenge energy in real time, from the patch of sky or air directly above them, at whatever rate the weather permits. And usually that's not much – at the height of a British summer you can spend all day outdoors and still be pale blue at supper time.

The third is net energy. It takes energy to get energy. Gas wells must be drilled; solar panels must be manufactured. The effective density of any source is reduced by the energy required to obtain it. For nuclear and conventional fossil fuels, that deduction is modest. For wind and solar, it is severe – the return falls to single digits, and the manufacturing energy comes overwhelmingly from the very hydrocarbon system they propose to replace. The net density gap is even wider than the gross one.

These three factors – source physics, time-space compression and net energy – are not policy choices. They are hard physical constraints, and no subsidy, no mandate and no amount of wishful thinking can override them. The obvious response is to build more of the dilute source. But that introduces a problem of its own, and to see it we need the third concept.

AREAL POWER DENSITY

Energy density tells you how much energy is stored in a kilogram of fuel. But a civilisation doesn't just need energy. It needs energy *at a certain rate*. The lights must stay on now, not next week. The steel furnace must run continuously, not just when the wind picks up. This introduces a new dimension: time. Power is energy delivered per unit of time. A concentrated source can deliver energy at a ferocious rate from a small site. A diffuse source cannot – and the only way to increase its rate of delivery is to collect it from more land.

'Areal power density' measures the amount of land needed to produce a specific amount of power: the rate of energy flow per unit of land area, in watts per square metre. The answer varies by a factor of a thousand, and that single fact explains more about the feasibility of energy policy than any amount of economic modelling.

Let's make this concrete. A gas-fired or nuclear power station generates roughly 1,000 watts from every square metre of land it occupies. A solar farm manages five to ten watts per square metre. A wind farm – once you account for the spacing that turbines need to avoid stealing each other's wind – delivers just one to three watts per square metre.[1]

Why does that matter? Think of the difference between mining a gold vein and panning a river. A gold vein is concentrated wealth: a single narrow shaft, a small patch of land, an enormous yield. That is a gas plant or a nuclear reactor – a ferocious rate of delivery from a tiny site.

Panning for gold is the opposite. The gold is real, but it is scattered in tiny flakes across miles of riverbed. To collect the same amount, you don't just need more pans – you need access roads along the river, processing equipment at every site and a huge workforce spread across the entire valley.

That is the power density problem. In a high-density system, the density does the heavy lifting – the energy is already where you need it. In a low-density system, you spend a massive fraction of the energy you will procure just building the collection network to go out and get it.

You've collected the same amount of gold, but the operational cost has eaten your profit.

The arithmetic is unforgiving (see Figure 2). To replace a single gas plant with wind turbines, you need somewhere between 300 and 1,000 times more land. And that land – outside of the Sahara and some other deserts – is generally not empty. It is farmland, moorland, coastal seabed or someone's horizon. In addition, the turbines

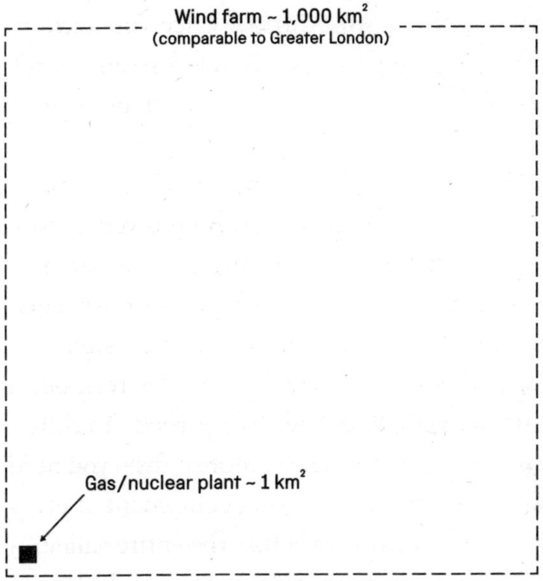

Fig.2 Land area required to generate one gigawatt of continuous power. The filled square is the gas plant (~one square kilometre). The dashed outline is the wind farm (~1,000 square kilometres, comparable to Greater London).

must be manufactured, transported, erected on concrete foundations, connected by access roads, and linked to the grid by hundreds of miles of new transmission lines. Every square metre of that sprawl has a material cost and an energy cost.

This is not a problem that improves with scale. It gets worse. When power density is low, scaling up means spreading out, and spreading out means longer transmission distances, greater infrastructure and more energy spent building and maintaining the collection network. As we'll see in a later section, at some point the energy required to build and sustain the system begins to consume a significant fraction of the energy it produces. The system is running to stand still.

The pattern is the same one we saw with energy density. Every successful energy transition in history moved towards *higher* areal power density. Burning wood in a hearth: perhaps 0.5 watts per square metre of the forest that supplies it. A coal mine feeding a power station: hundreds of watts per square metre. An oil well: higher still. A nuclear reactor: the most power from the smallest footprint humanity has ever achieved.

The direction has always been the same: *more energy from less land*. More capability from a smaller footprint. The proposal to replace gas and nuclear with wind and solar reverses this trajectory entirely. It is a step backwards to the kind of energy system that constrained civilisation for millennia – one that is spread thinly across the landscape, hostage to geography, weather and the brute mathematics of area.

That matters, because land is not abstract. It is finite, and it is already spoken for. An energy system that demands hundreds of times more of that land is not merely inconvenient. It is in direct competition with food production, with ecosystems and with the communities that live on the land it needs. The cheerful phrase 'energy transition' obscures what is actually being proposed: one of the largest and most destructive land-use changes in human history.

But before we can answer the question of what these three concepts mean, taken together, there is a distinction to make – and it is one that trips up much of the energy debate.

FUELS AND CARRIERS

A *fuel* is a store of energy that exists in nature: coal in a mine, oil in a reservoir, gas in a field, uranium in rock. You can dig it up, move it and release its energy when and where you choose. An *energy carrier* is different. It is a way of moving energy from one place to another. Electricity is the most familiar carrier. Unlike a fuel, it cannot be extracted from a deposit and stored – it must be manufactured, in real time, from a fuel, and every conversion loses energy.

When someone proposes to 'replace fossil fuels with electricity', they are not proposing a like-for-like swap.

Something still has to generate the electricity. They are proposing to replace a fuel with a carrier – and hoping nobody notices the difference.

THE THERMODYNAMIC FLOOR

Let's take stock. You now have three concepts – energy gradient, energy density and areal power density – and one critical distinction: a fuel stores energy, a carrier merely moves it. Between them they describe the physical constraints that every energy system must satisfy. No policy, no technology, and no amount of imagination, determination or investment can override them. They are the thermodynamic floor beneath civilisation.* Together, they define what this book will call energy *quality* – the gradient, the density and the areal power density that determine whether an energy source can sustain industrial civilisation or not.

The historical record bears this out. Wood to coal. Coal to oil. Each step moved up on all three measures simultaneously – steeper gradients, denser fuels, higher power density per unit of land – unlocking capabilities that the previous level could never have supported. Call it the quality ladder. The discovery of nuclear fission identified the next

* Thermodynamics is the branch of physics that governs how energy moves and changes form. The 'thermodynamic floor' is simply the set of physical rules that no energy system can break.

rung. Railways, aviation, the globalised supply chain. The arrow has always pointed in the same direction: towards higher quality. There are no successful counter-examples.

And here is the part that is not discussed in polite company. There are many unsuccessful counter-examples, involuntary moves *down* the quality ladder – deforestation, resource depletion, imperial overreach – that produced not a gentle simplification but a catastrophe. People die, in large numbers, when the energy that sustains their society contracts faster than they can adapt. The Western Roman Empire did not 'transition' to a lower-energy economy. It collapsed, and the population of its former heartlands – Italy, Spain, the Balkans – fell by roughly half over the centuries that followed.[2] That is what moving down the energy ladder looks like when it is not a choice.

The renewable energy transition talks – rightly – about the *quantity* of solar and wind energy available to us. It rarely talks about the poor *quality* of the sources, and what it costs in energy and resources to collect energy that arrives so thinly spread. You cannot solve a quality problem with quantity – that is what a thermodynamic floor means. And yet the current energy policy of most Western governments proposes to do exactly that: move down the quality ladder deliberately, replacing concentrated fossil and nuclear energy with whatever the wind and sun happen to deliver on any given afternoon.

The ambition is wrapped in the vocabulary of progress – 'clean energy transition', 'net zero', 'green growth'

– but the physics points in the opposite direction. A wind turbine is a lower-gradient, lower-density, lower-power-density device than the gas plant it proposes to replace. More euphemistic vocabulary conceals the consequences: 'degrowth' sounds like prudent management. The honest term is energy rationing, and energy rationing – 'demand-side flexibility' in the argot of the advocates – is never a choice made by prosperous societies. It is a condition imposed by crisis.

When that crisis is not resolved, the result is always the same: declining living standards, eroding social stability and – eventually – war over the resources that remain. The question is not whether a society based on low-quality energy will be unpleasant. The question is whether it will be survivable at the scale of eight billion people. The physics suggests that it is not.

But before we consider what can be done about this, we need to understand what all this energy is actually *for*. The answer is not 'electricity'. The answer is the four industrial processes without which modern civilisation cannot exist. That is the subject of Part II.

Part II
The Industrial Metabolism

In which we discover what civilisation actually eats, and why you can't replace the menu with electricity.

That was the physics. This is the chemistry. And the chemistry is worse.

The centrepiece of current energy policy is a single proposition: replace fossil fuel combustion with electricity generated from wind and solar. Electrify heating. Electrify transport. Electrify industry. The shorthand is 'electrify everything', and it is the organising principle behind net zero, green growth and the clean energy transition.

To understand why this will not work, you need to know what industrial civilisation actually *consumes* – not in the abstract, but at the level of specific processes at specific temperatures. You need to understand civilisation's *metabolism*.

A metabolism is the set of chemical processes by which a living thing converts raw materials into the substances it needs to survive. Your body has one. It takes in specific inputs – protein, carbohydrate, fat, water – in specific chemical forms and in specific quantities, transforms them and excretes what it cannot use. You cannot substitute one input for another. You cannot replace protein with extra water and expect to survive. You cannot live on air and good intentions, however fervently you believe otherwise. If any part of the system fails – input, conversion or output – you die. Not eventually. Quite soon.

Industrial civilisation is not a living thing, but the analogy is exact. It too has a metabolism – a set of chemical processes that convert raw materials into the substances on which its survival depends. It consumes raw materials, transforms them at extreme temperatures using specific chemical reactions, and produces the physical substances on which everything else depends. The four most important of these products – the pillars that hold up civilisation – are steel, cement, ammonia and plastics.

THE FOUR PILLARS

Steel holds up almost every building you have ever entered, almost every bridge you have ever crossed, almost every vehicle you have ever ridden in. Making it requires heating iron ore to around 1,500°C in a blast furnace, using coke – derived from coal – as both the heat source and the chemical agent that strips oxygen from the ore. The carbon is not optional.* Without it, the iron stays as rock. Global production: roughly 1.9 billion tonnes a year.[3] Without steel, there are no cities, no railways, no ships, no surgical instruments, no wind turbines.

Cement is the binding agent in concrete, which is the most widely used material on Earth after water. Making cement requires heating limestone and clay to about 1,450°C in a rotating industrial furnace. The calcium carbonate in the limestone releases carbon dioxide as an

* Experimental hydrogen-based routes exist but require vast quantities of clean hydrogen that do not yet exist at scale – a problem explored in Part v.

intrinsic part of the chemical reaction – not because the kiln burns fossil fuel (though it does), but because the chemistry *itself* produces CO_2. Even if you powered the kiln with clean electricity, roughly two-thirds of the CO_2 emissions would remain.[4] Global production: roughly four billion tonnes a year.[5] Without it, there are no foundations, no dams, no roads, no runways.

Ammonia is the basis of virtually all synthetic fertiliser, which in turn sustains roughly half the world's food supply. Making ammonia involves a chemical process that strips hydrogen from natural gas and combines it with nitrogen from the air at 400–500°C under extreme pressure. The natural gas provides both the hydrogen feedstock and the energy to drive the reaction. Global production: nearly 200 million tonnes a year. Without it, crop yields collapse and billions face starvation.[6]

Plastics are built from hydrocarbon chains derived almost entirely from oil and natural gas. The fossil fuel is not consumed. It is *restructured*, molecule by molecule, into materials that are lighter than metal, cheaper than glass and more versatile than either. Global production: over 400 million tonnes a year.[7] Without them, there is no modern medicine, no food packaging, no insulation, no electronics.

Four materials. Four processes that require specific chemical inputs at extreme temperatures. Between them, they account for roughly a third of global energy use.[8] Four dependencies on hydrocarbons not as a *fuel* but as a *feedstock*. Remove any one and the system that feeds, houses and sustains eight billion people begins to fail. Remove all four and it ceases to exist. So what happens when someone proposes to replace these feedstocks with electricity?

THE ELECTRIC DIET

There is a community of people who believe that humans can live on air and sunlight. They call themselves breatharians. They hold conferences, publish books and maintain a sincere conviction that the human body does not require food. Periodically, one of them dies.[9]

The proposal to run an industrial civilisation on electricity alone is the breatharian diet of energy policy. It begins with a grain of truth – electricity is marvellous for many things – and extends it into a fantasy: if we simply generate enough clean electricity, we can replace every industrial process that currently depends on hydrocarbons. 'Electrify everything' is a slogan, not an engineering plan.

Consider what 'electrify everything' must actually mean for the four pillars. To make steel without coal, you need hydrogen – vast quantities of it – produced by splitting water with electricity.* To make cement without releasing

* Hydrogen (H_2) is a gas that releases energy without CO_2 at the point of use, which is why it is proposed as a fossil-fuel substitute. The catch is in producing it. Today, almost all industrial hydrogen is made by reforming natural gas – itself a fossil fuel process. 'Clean' hydrogen requires splitting water using clean electricity. The scale required is examined in Part v.

process CO_2, you would need to reinvent the chemistry of calcium carbonate, a problem no one has solved at scale. To make ammonia without natural gas, you would need the same clean hydrogen in even vaster quantities. To make plastics without oil, you need synthetic hydrocarbon feedstocks built up from CO_2 and hydrogen – a process so energy-intensive that it makes the original fossil route look efficient by comparison.

Every one of these alternatives requires enormous amounts of cheap, reliable, high-density, high-gradient electricity that does not yet exist. And every one of them runs into the distinction we established in Part I: electricity is not a fuel. It is an energy carrier – manufactured, in real time, from a primary source, with losses at every step. You cannot mine electricity. You cannot store it in a barrel. You can store it in a battery or convert it to hydrogen, but each conversion loses energy and adds cost – and the result is still a carrier, not a fuel. When someone says, 'Replace fossil fuels with electricity', they are not proposing a like-for-like substitution. They are proposing to replace a fuel with a carrier – and then asking that carrier to do the chemical work that only a fuel can do.

And here is the circularity that its advocates prefer not to discuss: building the wind farms, solar arrays, battery storage and transmission lines requires steel, cement and plastics – which can only be produced using hydrocarbons. And because these systems wear out and must be rebuilt every twenty to thirty years, the dependency is not temporary. It is permanent. You do not use the old system to launch the new one and then switch it off. You use the old system to *sustain* the new one, indefinitely.

The honest version of 'electrify everything' is therefore this: electrify the things that electricity can actually do – lighting, computing, rail transport, heat pumps – and accept that the industrial core of civilisation will continue to require hydrocarbons as a chemical input for the foreseeable future. That is not a failure of ambition. It is an acknowledgement of chemistry.

But before examining what happens when we try to build the replacement anyway, there is a prior question that most energy debates skip entirely: what is happening to the fossil fuels themselves? The answer, as Part III reveals, makes everything that follows considerably more urgent.

Part III

The Depleting Inheritance

In which we discover that the one-time endowment of fossil sunlight is running out faster than anyone in authority will admit.

Parts I and II described what energy is, what it does and why the industrial metabolism depends on hydrocarbons not merely as fuel but as feedstock. Before asking whether renewables can replace them, there is a prior question that the energy debate has quietly buried: what is happening to the fossil fuels themselves?

You may have heard that the world is running out of oil – that oil has 'peaked'. You may also have heard that this fear was debunked – that new technology keeps finding more, that production keeps rising, that the doomsayers were wrong. Both claims contain a grain of truth, and the gap between them is where the danger lies. The answer turns on a single distinction: not all oil is equal. Once that distinction is clear, the comfortable story falls apart.

PEAK ECONOMIC OIL

Not all oil is equal. This is the fact that the 'peak oil was wrong' narrative has never understood.

A barrel of conventional crude from Saudi Arabia costs roughly $10 to extract and returns over thirty times the energy invested. A barrel from the Canadian oil sands is not really oil at all – it is bitumen, natural tar, so heavy it must be mined or melted out of the ground with steam, then upgraded with vast quantities of natural gas before a refinery can even process it. The break-even cost is roughly

\$30 to \$40 a barrel, and the energy return is only around 4:1 – against 30:1 for Saudi conventional. A barrel from a US shale well costs somewhere in between, with the well declining violently after the first year, which must be replaced by continuous drilling. All three appear in the reserves statistics as 'oil'. But they are not the same resource in any sense that matters to an industrial civilisation. One sustains the system. The others barely sustain themselves.

There is an oil price above which the economy contracts. That threshold is hard to pin down precisely, but its existence creates an invisible partition in our remaining reserves. On one side is economic oil – the volume the world can afford to burn. On the other is oil so costly or energy-intensive that it is effectively non-existent.

But the line between them is not fixed; it is a moving target determined by a global blend. The economy can absorb expensive, difficult oil only as long as a massive base of cheap, easy oil dilutes the total cost. A \$120 barrel is viable if the volume-weighted average is \$60.

The danger is not that any single barrel becomes too expensive, but that our buffer of cheap, conventional crude is shrinking. As that cheap base pulls back, the expensive sources it was carrying drag the global average upwards. At some point the average cost exceeds what the economy can bear. Growth stalls. Demand falls – and with it, the price. But when the price drops, the expensive oil that was only marginally viable at the higher price is no longer viable at all. It stops being produced – not gradually, but suddenly. Production lurches down, and the economy – starved of the energy that production was supplying – contracts again, dragging demand and price lower still and killing more marginal supply. This is not a smooth decline. It is a feedback loop – and it is invisible right up until it triggers, because the blended average masks the fragility underneath. The signature of depletion is therefore not a smoothly rising price but violent oscillation – spikes that break the economy, busts that destroy supply, each cycle starting from a lower base. How that grinding contraction becomes inflation – through the money printed to fight it – is the story of Part VI.

The optimist's response is usually to point to 'proven reserves' – the 1.7 trillion barrels that appear in official statistics – and declare that the world has decades of supply remaining. But proven reserves are not what they appear to be. The label has a strict technical meaning – oil recoverable profitably, with reasonable certainty, at today's prices and costs – and the headline figure, an unaudited tally of national declarations, has never been held to it. It includes oil in jurisdictions too unstable to develop. It includes oil so deep, so heavy or so tightly locked in rock that extracting it may consume nearly as much energy as it yields. And they include oil that is purely imaginary* – OPEC members

* OPEC member states self-report reserves without independent audit. The mid-1980s additions are documented in Campbell and Laherrère, 'The end of cheap oil', *Scientific American*, March 1998.

inflated their stated reserves by hundreds of billions of barrels in the 1980s, when quota allocations were tied to reserve size, with no corresponding new discoveries. The number that matters is not how much oil is in the ground. It is how much can be extracted at a rate, cost and net energy return sufficient to keep civilisation running. That number is considerably smaller – and it is shrinking.

'Peak oil' always meant peak economic oil – though the early thinkers never needed to say so, because in their day virtually all oil was economic. Tar sands and tight shale were geological curiosities, not production targets. The claim was that the cheap, high-quality oil on which industrial growth depends would peak and decline, and that what remained would be progressively harder to extract, more expensive and less productive. That is exactly what happened. The distinction only became necessary when central banks flooded the financial system with imaginary money and debt, and that money found its way into sources the market had already rejected.

Economic oil – the cheap conventional oil that built the modern world – peaked around 2006.* The International Energy Agency's 'World Energy Outlook 2010' acknowledged that conventional crude production had reached

* Definitions of 'economic' vary between sources, and with them the precise peak year, but the discovery–production gap shown in Figure 3 makes the trajectory plain. By any reasonable definition, cheap oil has peaked and is in decline.

its maximum and would never return to that level.[10] The event passed almost without public comment. It should have been the most important news story of the century.

What happened instead was the 2008 global financial crisis, an event universally attributed to reckless lending, but which makes considerably more sense when understood as the first collision between a growing economy and a contracting energy supply – and, as we shall see in Part VI, not the last. Oil prices hit $147 a barrel in July 2008. Within months, the financial system broke. The causal chain – energy constraint, price spike, collapsing demand, financial contagion – is visible in retrospect, but almost no one in authority saw it.

THE $200 TRILLION FIX

The world did not consciously choose to ignore the constraint – almost no one in authority recognised it as such. But the response amounted to a choice nonetheless: rather than investigate why oil had hit $147 a barrel, central banks treated the crisis as a purely financial event and reached for the only tool they had. The price had spiked because conventional crude production had stopped growing – and has been stuck on a plateau ever since 2005. The geological constraint that M. King Hubbert had warned of had finally bitten. Hubbert, an American geophysicist working for Shell, had shown in the 1950s that the output of a conventional oil province follows a bell-shaped curve

– rising to a peak as the easy reserves are found, then declining however hard we drill. He predicted that conventional US production would peak around 1970, and it did.

That tool was imaginary money. Between 2008 and 2022, the US Federal Reserve and its G10 counterparts expanded their combined balance sheets from roughly $7 trillion to nearly $35 trillion – conjuring some $28 trillion into existence through quantitative easing and emergency asset purchases.[11]

What followed was extraordinary. With interest rates held near zero for over a decade, borrowing became almost free – and governments, corporations and households around the world obliged. Since 2008, global debt has nearly doubled, from roughly $175 trillion to a record $348 trillion.[12] Every dollar of it is a claim on future economic output, and therefore on future energy. (We shall return to that thought in Part VI.) A significant share of the American portion flowed into the shale oil industry – an industry that had existed in experimental form for decades but had never been commercially viable, because the oil it targeted was too tightly locked in rock to extract at a profit.

Cheap capital changed the arithmetic. With borrowing costs near zero, companies could drill at a loss and refinance indefinitely. US oil production – which had been in long-term decline since 1970 – tripled from roughly five million barrels a day in 2008 to over thirteen million by 2024.[13] America became the world's largest oil producer.

The shale revolution was hailed as proof that human ingenuity had once again overcome physical limits.

But the oil that shale produced was not the same as the oil it replaced. Conventional crude flows from a reservoir under its own pressure; a well can produce for decades. A shale well must be hydraulically fractured – rock shattered at high pressure to release oil trapped in microscopic pores – and its production drops by 25 to 40 per cent in the first year. A conventional field declines gently over a generation. A shale well declines violently in months. To maintain output, you must drill continuously. To grow output, you must drill exponentially. The industry did not discover a new oil province. It built a treadmill.

The financial record confirms this. The US shale industry chronically spent more than it earned for over a decade. It survived not on the revenue from the oil it produced, but on the continuous injection of new capital from investors chasing a growth story that the physics could not sustain. The industry did eventually turn cash-flow-positive after 2020 – but only because the price crash that year forced a wave of bankruptcies that wrote off the bad debts, the Federal Reserve bought oil-company bonds for the first time in its history to stabilise the survivors and those survivors slashed their drilling budgets to harvest existing wells rather than develop new ones.

By the mid-2020s, the bill was coming due. The best acreage – high-yield locations in the Permian Basin, the Bakken, the Eagle Ford – was being exhausted. In 2024, for the first time, productivity per lateral foot of drilling declined. The industry was drilling deeper wells to extract less oil. The treadmill was accelerating, and the runner was slowing down.

So the claim that 'peak oil was wrong' amounts to this: we created $28 trillion of imaginary money, piled up $173 trillion in debt on top of it, and used a fraction of the proceeds to mobilise a vast volume of uneconomic oil. Because total production kept rising, we declared the problem solved. But the physics did not change. The accounting did. And the time we bought was borrowed from the future – because every barrel of uneconomic oil extracted today is a barrel unavailable tomorrow, when, as we shall see, the need will be greater and the alternatives fewer.

But shale is only the most visible part of a deeper problem. The violent decline of a shale well is an extreme case of something that happens to every oil field on Earth. To see why the crisis is not merely American, we need to understand the physics that governs them all.

THE PHYSICS OF DEPLETION

An oil well is not a tank with a tap. When you drill into a reservoir, the oil flows as a result of natural, stored energy – pressure from the aquifer beneath, expansion of gas dissolved in the oil, the weight of the rock itself. As you extract oil, that stored energy depletes. The pressure drops. The flow slows. Eventually you need pumps, then water injection, then gas injection, each consuming more energy and recovering less oil. Every well follows the same fundamental arc: maximum flow at the start, then decline. Reworking existing wells and applying new techniques can slow the decline or buy temporary recoveries, but the underlying trajectory is set by the depletion of stored energy in the reservoir. That cannot be reversed.

To see why this matters, imagine a row of spinning plates on sticks – like a circus act. Each plate is an oil well. New plates are being set spinning all the time, but older plates are always slowing down and falling. To keep the same number spinning, you must start new plates at least as fast as old ones fall. To increase the total, you must start them faster. This is the fundamental dynamic of oil production: it is not a stock you draw down at will. It is a declining flow you must constantly renew.

The International Energy Agency quantified this in 2025. If all investment in existing oil fields were to stop tomorrow, global production would fall by 8 per cent per year.[14] That may not sound much, but at 8 per cent, global oil production would halve in under nine years.* Then halve again in the next nine years.

This is the treadmill that the oil industry now runs on every day. Nearly 90 per cent of all investment in finding and producing oil now goes simply to offsetting decline in existing fields – running to stand still. The remaining 10

* To get the halving time, apply the 'Rule of 70' – divide 70 by the percentage decline rate.

per cent is all that is available to grow production. And the treadmill is accelerating: as the world's giant fields age and newer, smaller, faster-declining sources like shale take a larger share, the decline rate steepens.

There is a subtlety here that is easy to miss. Investment that increases the rate of production from existing economic reserves – running fields harder, drilling additional wells between the existing ones, injecting more water – does not add oil. It pulls it forward in time. The total recoverable volume does not change. You get more today and less tomorrow. The peak arrives later, but the decline on the far side is steeper – because the easy oil has been extracted faster, leaving less to cushion the fall. There is a rate of decline below which an economy can adapt: industries retool, alternatives are built, demand adjusts. Above that rate, the adjustment outruns the economy's capacity to respond. The difference is between a managed descent and a crash. Every barrel pulled forwards narrows the margin.

So the treadmill demands a constant supply of new fields to replace the ones that are dying. The critical question is: where do new fields come from?

DISCOVERY CONSTRAINS PRODUCTION

They come from discovery. Since you cannot produce what you have not first discovered, the pattern of discovery determines the pattern of production. This is the most basic fact in the oil industry, and the one most consistently ignored in energy policy.

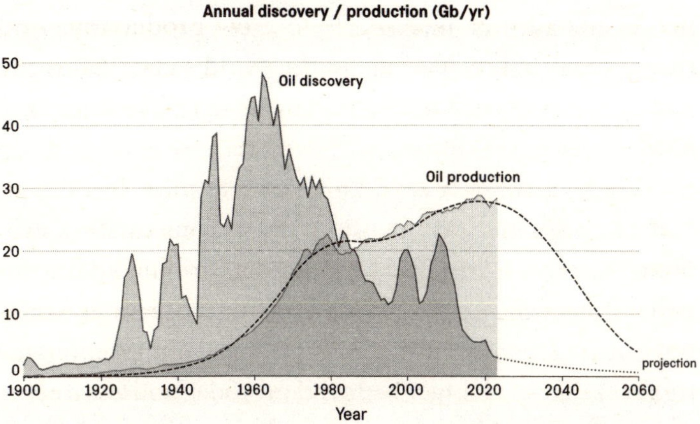

Fig. 3 Global conventional oil discovery and production (excluding ultra-heavy), after Campbell and Laherrère, 'The end of cheap oil', *Scientific American*, 1998, updated with J. Laherrère's 2024 data. Discovery volumes peaked in the 1960s and have declined despite massive improvements in exploration technology. Production continued to rise, drawing down the finite volume already discovered. The widening gap between the two curves represents the accelerating consumption of a one-time inheritance.

Global oil discovery peaked in the 1960s (see Figure 3), when the industry was finding roughly fifty billion barrels a year.[15] That era and the decade that followed opened the great provinces – the Middle East super-giants, western Siberia, Alaska, the North Sea. The tools have improved enormously since then: seismic imaging, deep-water drilling, horizontal wells, machine learning applied to geological data. The planet has been surveyed with a thoroughness that would have astonished a 1960s geologist. And yet discoveries have fallen, decade by decade. In 2024, the world discovered under two billion barrels while consuming thirty billion – burning through our inheritance more than fifteen times faster than we were finding new deposits.

This pattern of collapsing discovery reveals something else – something that matters even more than the annual shortfall. It can be used in a simple but powerful statistical technique that allows geologists to estimate the total volume of oil the Earth will ever yield. Add up every barrel discovered, year by year, as a running total. Once annual discoveries begin falling after the 1960s peak, the running total begins to slow – and its ultimate destination becomes possible to project. With each subsequent decade of diminishing discovery, the projection converges. The uncertainty shrinks. The geologists' term for the destination is *ultimate recoverable resource*, or URR.

Today that number is essentially known. Barring an astonishing technological breakthrough – a possibility we address later – the cumulative discovery curve has almost stopped rising and can only flatten further. The URR is not a number we are guessing at; we can already read it off the curve. That number is approximately two trillion barrels of conventional crude. About half has already been produced. Higher figures appear in official agency assessments, which add generous estimates of yet-to-be-found oil and unconventional resources – but the discovery curve itself, drawn from actual finds backdated to the year each field was found, is unambiguous.

Now apply this to production. You cannot pump what was never discovered. Production is capped by the URR. And the URR is already known. The peak of global oil production is not a matter of opinion, or modelling, or politics. It is a matter of arithmetic – and the arithmetic is already done.

TECHNOLOGY CANNOT CLOSE THE GAP

The instinctive response is: 'But technology will find more.' It is worth pausing on why this is wrong.

Peak discovery – the 1960s – coincided with the Apollo programme. Humanity was landing on the Moon. In the six decades since, we have lived through the greatest technological revolution in history: electronics, miniaturisation, computing, satellite surveillance, materials science, artificial intelligence. Every one of these has been applied to oil exploration. The result? Discovery volumes fell by over 90 per cent. Production of economic oil followed the predicted path.

The largest fields are found first, because they are the easiest to find, and the easiest are produced first. The maths that governs what happens next is simple and brutal, and the pattern it predicts is exactly the one the data show. Technology does not defeat this maths. It confirms it. Sixty years of the most sophisticated exploration in history have found what remains: smaller, deeper, faster-declining deposits – plates that spin down almost as soon as they start. But the treadmill needs giant fields to replace the giants that are dying. There are none left to find.

THE 'NEW SAUDI ARABIA' TEST

If you still aren't convinced, try a thought experiment. Suppose tomorrow we discovered a single province containing 260 billion barrels – equivalent to Saudi Arabia's entire claimed reserves, which took decades and dozens of fields to accumulate. How much difference would it make?

At current global consumption of roughly thirty-seven billion barrels a year, a new Saudi Arabia would supply seven years of demand. A discovery of unprecedented scale would delay the peak by a few years at most. It would not prevent it.

The physics of depletion, the pattern of discovery, the maths of easy-first-hard-last development and the limits of technology all point to the same conclusion. The question is no longer whether global oil production will peak. It is how much of the remaining endowment is actually usable – and whether it is enough to provide us with a bridge to whatever comes next.

THE LEDGER

Let us draw up the account.

Rystad Energy – the industry's most widely cited independent data source – estimates that at current production rates, the world's proven oil reserves will last roughly fourteen years.[16] That does not mean the oil stops in fourteen years. It is a snapshot – today's proven inventory at today's burn rate, with no new oil added to the books. New reserves are added each year as fields are appraised and prices shift.

But the rate of addition has been falling for decades. The oil majors know this. They have recently spent over $110 billion buying reserves from each other – ExxonMobil acquiring Pioneer, Chevron acquiring Hess – because they cannot find new ones. But acquiring oil from a rival does not create new oil. It reshuffles a shrinking deck. In 2024, the global industry discovered 1.8 billion barrels while producing thirty billion – a replacement rate of 6 per cent.

Fourteen years is not a generation. It is not a planning horizon. It is the time it takes to design, approve, finance and build the replacement for a single power station.

Gas is in a different position. Proven reserves stand at roughly fifty years of current consumption. This is a significant buffer – but it is not a solution to the oil problem, because oil and gas are not interchangeable. Oil is a liquid. It powers virtually all transport: almost every car and truck, every ship, every aircraft. It is the feedstock for plastics, pharmaceuticals and the entire petrochemical supply chain described in Part II. Gas can heat buildings, generate electricity and provide some petrochemical feedstock, but it cannot fuel a truck fleet, fly an aircraft or replace naphtha – the petroleum feedstock on which the chemical industry depends – in the steam crackers that produce plastics, pharmaceuticals and most industrial chemicals. The depletion of oil is therefore not softened by the abundance of gas. It is a separate crisis, running on its own clock.

Now put the ledger together with what this book has argued so far.

The hydrocarbon endowment is finite and depleting. Oil – the most energy-dense, most versatile, most irreplaceable fraction – is depleting fastest. The shale response bought time but masked the decline of the economic reserves on which civilisation depends. Worse, every technique that accelerates production without adding reserves – running fields harder, drilling faster – pulls oil forwards in time, steepening the decline that follows. And the endowment we are depleting is the same endowment we would need to build *any* replacement – including one that might actually work.

This is the context in which every proposed energy transition must be judged. It is not enough for a replacement to 'work' in laboratory conditions or in a spreadsheet. It must work while the inheritance that funds it is shrinking. Every barrel of the remaining endowment matters. Every barrel spent on a renewable energy system that requires permanent hydrocarbon backup – gas plants on standby for when the weather fails – and must be rebuilt every generation using more hydrocarbons, is a barrel unavailable for the bridge to something that can actually sustain itself.

There will come a point – and it is closer than anyone in authority admits – when the remaining oil is insufficient both to maintain the existing system and to build a new one to replace it. At this point, there will be fifty miles of fuel in the tank, with sixty miles to the nearest petrol station.

The next two parts examine the replacements we have been promised – and whether any of them can work before the tank runs dry.

Part IV

The Renewables Paradox

In which we discover that renewables aren't renewable, and that building the replacement for fossil fuels requires more fossil fuel than it saves.

It's not what an energy system produces that sustains civilisation. It's what the energy system has to spare after it has fed itself – the *net* output. Every energy source consumes energy – to drill the well, mine the ore, build the infrastructure and deliver the output. What matters is how much is left over. For most of the fossil fuel era, the distinction between gross and net has barely mattered.

For wind turbines and solar panels, the energy tax is not trivial. They must be mined, manufactured, installed, connected, backed with storage and replaced when they wear out – and every one of those steps consumes energy. The surplus that remains is far smaller than advocates suggest – and it gets worse the harder you look.

Part IV examines how much worse in five steps. First, the bridge fuel problem – why a transition that abandons its high-quality energy sources cannot generate the surplus energy needed to replace them. Second, the energy cliff – the hidden mathematics of energy return can make a small decline in efficiency catastrophic. Third, the storage problem – what happens when you try to make an intermittent source do the job of a continuous one, and what it actually costs. Fourth, the physical bill of materials – the steel, concrete, copper and lithium that the buildout demands, and where it all comes from. Fifth, the replacement treadmill – what happens when a system with a twenty-five-year lifespan built from non-renewable resources must be rebuilt, again and again, forever.

THE BRIDGE FUEL PROBLEM

Every energy transition in history has been like changing an aircraft's engines in flight, and ours is no different. The new system must be built while the old one keeps running – because you cannot stop feeding, heating and transporting the world's eight billion people while the work is done. Four billion of them depend on food grown with fossil fuel fertiliser, for example. They cannot wait.

This is the bridge fuel problem: the energy needed to bridge from the present system to the future one while keeping everything running. To provide that bridge, every successful transition has met two conditions.[17] First, there was a large surplus of the current energy source still available. Second, the new source was higher quality – more energy from less material and less land. Wood to coal, and coal to oil, both met both conditions.

The proposed transition violates both conditions. As we have seen, our current energy source is running out – fast. And we are being asked to use what's left to build lower-quality energy infrastructure, hoping the new system can eventually sustain itself. But if the new system returns less net energy than the fossil fuels used to build it, you can neither maintain current civilisation nor construct its replacement. Whether the proposed renewable transition meets that test is the question the rest of this part examines.

History has another name for this failure: collapse. It is what happened when the Western Roman Empire exhausted its accessible fuel supply – deforestation was one of several factors that forced it down the energy ladder, and the population of Rome fell from over a million to perhaps 30,000.

How far down the energy ladder does the proposed transition push us? The answer is hidden in a ratio that most people have never heard of – and in a cliff that makes the arithmetic merciless.

THE ENERGY CLIFF PROBLEM

The ratio between the energy you get from a source and the energy you spend obtaining it is called EROEI – Energy Return on Energy Invested (see Figure 4). It is the defining ratio of any energy system, because it determines how much useful energy is left over after building, operating, maintaining and replacing your energy system to do what actually matters: to heat homes, to run factories, to grow food, to sustain a civilisation.

At an EROEI of 30:1, you invest one unit of energy and get thirty back. Twenty-nine are available for society. The energy cost of obtaining your energy is modest – about 3 per cent of the total. This is roughly where conventional oil and nuclear have sat for most of the

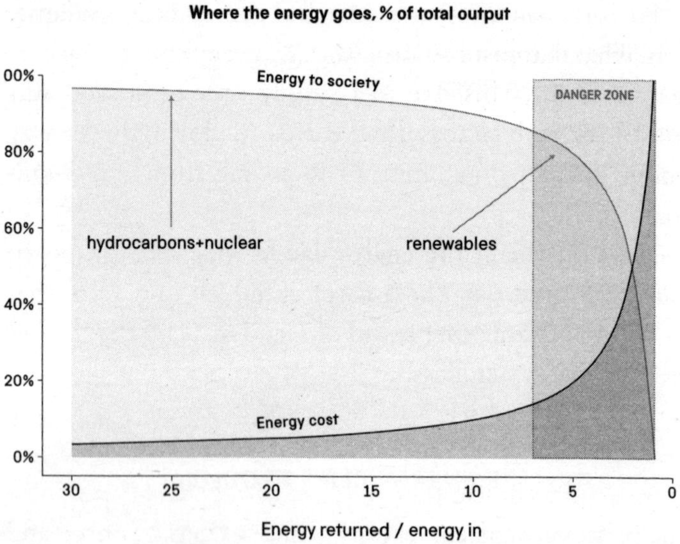

Fig.4 The energy cliff. As EROEI declines from left to right, the share of total output consumed by the energy system itself (dark shading) balloons, causing the share available to society (light shading) to collapse. Hydrocarbons and nuclear sit comfortably on the left. Wind and solar, once storage and grid costs are included, sit in the danger zone – where history shows us societies cannot function.

energy returned by the individual *device* – the wind turbine, the solar panel, etc. – not by the *system* needed to make the device compatible with the demands of civilisation. The difference is what it takes to make the devices reliable. And the difference is catastrophic.

What system? A wind turbine in isolation may return sixteen units of energy for every unit invested. But a wind turbine in isolation generates intermittent energy, and civilisation tolerates an intermittent energy supply about as well as you tolerate an intermittent air supply. You don't want your baby in a neonatal unit that works occasionally. An energy system must deliver power reliably, and on demand.

That means either an energy system that is always on (such as nuclear), or storage to balance supply across time – for example batteries, pumped water storage or hydrogen, each of which consumes energy to build and loses energy in every charge-discharge cycle. It means grid infrastructure to balance supply across geography. It means backup generation for the weeks when the wind does not blow and the sun does not shine. Each addition subtracts from the system EROEI, pushing the real number further down the cliff.

The above are abstract claims until you test them against a real country. Of these, storage is the largest and most revealing – because it is the one you can test against real weather data. How much is needed? The United Kingdom is a useful case: a large, wealthy, wind-rich island that has committed by law to net zero by 2050. If the storage arithmetic works anywhere in the temperate world, it should work here.

A 2023 Royal Society study modelled thirty-seven years of UK weather data against projected electricity demand over the next thirty years and arrived at a sobering answer.[19] The problem is not merely the *dunkelflaute* – the dark doldrums of a windless winter week. The data reveal that UK wind output can remain persistently below its long-run median for two to three consecutive *years*, producing cumulative energy deficits that dwarf anything a battery can absorb. Meeting demand through a multi-year wind drought requires tens of terawatt-hours of stored energy – energy that must somehow have been deposited years earlier during an equally sustained surplus.

The United Kingdom currently possesses about twenty-four gigawatt-hours of pumped hydro storage and roughly thirteen gigawatt-hours of grid-scale batteries. The required storage is somewhere between sixty and one hundred terawatt-hours – that is, between 60,000 and 100,000 gigawatt-hours. The country has built less than one part in 1,600 of what is needed. Could it build the rest? Pumped hydro is the cheapest and most mature form of grid storage, but it requires specific geography – paired reservoirs at different elevations, connected by a tunnel. There are eleven schemes currently in the development pipeline across Scotland and Wales. If every one of them were built – and pumped hydro projects routinely take a decade or more – they would add roughly 200 gigawatt-hours, taking the UK total to around 240 gigawatt-hours. That is less than half of 1 per cent of what is needed. The geography for hundreds more does not exist. Lithium-ion batteries are the alternative, but at current installed costs of £100–250 per kilowatt-hour, building one hundred terawatt-hours of battery storage would cost £10–25 trillion – three to eight times the entire annual GDP of the United Kingdom – spent on batteries alone. And those batteries degrade and must be replaced every ten to fifteen years. And this is for electricity only, which accounts for roughly one fifth of UK final energy demand. Electrify heating, transport and industry, as current policy assumes, and the numbers become farcical.

And meeting these storage needs is not feasible – for two reasons. The first is technical. No battery technology stores electricity at continental scale for years. Lithium-ion batteries – the only type currently deployed at grid scale – self-discharge a few per cent every month, lose capacity to calendar ageing and have a service life of ten to fifteen years. They are designed for daily cycling, not for holding the output of one year through to the next. The system has to handle multi-year shortfalls: the Royal Society's analysis identified the UK's 2009–11 wind lull – three winters of persistently low wind speeds – as the period that drove its storage requirement to its maximum.

The second reason is economic. Even if such batteries existed, building them at this scale would not be affordable. The Royal Society investigated hydrogen as an alternative, but even that route was costed in the hundreds of billions of pounds, with the underlying geology limited to pockets of Cheshire, Wessex and East Yorkshire. That is why you also need backup generation – gas turbines, most likely, as they are the easiest to restart – capable of meeting the full demand of the grid. These must be built, maintained, staffed and connected, even though they will sit idle much of the time. In a system dominated by wind and solar, the backup fleet runs at a low capacity factor – meaning it generates only a small fraction of the time – but must still be paid for in full. The electricity it produces, on the rare occasions it is called upon, is ruinously expensive. You have not replaced the fossil fuel system. You have built the renewable system on top of it, and you are paying for both – not as a transitional cost, but as a permanent one, because the backup can never be retired.

Now recall the energy cliff from the previous section. A wind turbine's claimed EROEI of 16:1 measures the device alone. Add the storage, grid and backup needed to make it a reliable power source, and the system EROEI falls sharply. How far? No peer-reviewed study has established the energy return of a fully buffered wind-and-solar system at grid scale, because no such system has ever been built. Weißbach et al. estimated roughly 4:1, assuming pumped hydro, the most efficient storage available.[20] Barnhart et al., working at Stanford, showed that pairing wind with lithium-ion batteries actually reduces net energy return below what you would get by simply throwing the surplus away.[21] The reason: batteries cost energy to manufacture, lose energy on every charge-discharge cycle and must be replaced every ten to fifteen years. The energy spent building and rebuilding them exceeds the energy recovered from them. Palmer demonstrated that there is no generally agreed methodology for incorporating storage into EROEI at all.[22] The estimates that exist range from roughly 3:1 to 10:1, depending almost entirely on assumptions about storage – squarely on the part of the cliff where small errors can wipe out civilisation. We are proposing to gamble civilisation on a number no one has measured.

Whatever your reason for wanting an energy transition, a transition whose net energy return cannot sustain the civilisation it is supposed to power is not a solution to anything. It is the most expensive and elaborate method ever devised for committing civilisational suicide.

These are the energy numbers. Now translate them into physical reality – the steel, concrete, copper and lithium that the buildout actually demands.

THE BILL OF MATERIALS PROBLEM

The energy cliff is an abstraction until you translate it into physical reality. So let's do that now.

The International Energy Agency has calculated that an onshore wind plant requires nine times more critical mineral inputs than a gas-fired plant of equivalent capacity – and offshore wind thirteen times more.[23] The batteries required to back either source add lithium, cobalt, nickel and manganese in quantities that increase with every hour of storage demanded. These are large numbers. But they are only the *visible* part of the bill. The invisible part is vastly larger still.

Consider what it takes to put a single wind turbine on a hilltop. The steel in the tower was smelted from iron ore in a blast furnace burning coke. The ore was mined by diesel-powered excavators, transported by diesel trucks to a rail head, carried by diesel locomotives to a port and shipped across an ocean by a vessel burning heavy fuel oil. The concrete in the foundation was mixed from cement fired in a gas-burning kiln, from limestone quarried by diesel machinery. The copper in the generator was mined in Chile or the DRC, refined in a fossil-fuelled smelter, drawn into wire in a factory powered by a coal-fired grid. The rare earth magnets were extracted from ore in Inner Mongolia using acids produced in a petrochemical plant. The fibreglass blades are made from resins derived from oil. The lubricant in the gearbox is a petroleum product. The road built to reach the site was paved with tar from a refinery. The crane that lifted the aerodynamic shell into place runs on diesel. The vessel that installed the subsea cable burns marine fuel oil. The computer on which the maintenance contract was drafted is housed in a plastic casing derived from natural gas.

Now consider something that should stop you in your tracks. Every *item of equipment* that manufactured the turbine – the blast furnace, the excavator, the ship, the crane, the smelter – was itself manufactured by the same process, from the same inputs. And every item that manufactured *them*. The dependency is recursive, and it reaches all the way down. It's not a line. It is a web, and every strand in it leads back to hydrocarbons.

This is what we call the Global Industrial Manufacturing System. Every link in that chain is a hydrocarbon operation. Not some of them. Not most of them. *Every single one*. The supply chain that builds the 'clean energy system' is, from the mine face to the turbine blade, a fossil fuel enterprise. This is not a transitional inconvenience that will resolve itself as the buildout matures. It is the structure of industrial civilisation, and electricity cannot replace it – because, as we saw in Part II, the processes at every link require hydrocarbons not as a fuel but as a feedstock.

The International Energy Agency estimates that its own clean-energy scenario would require, by 2040, over forty times the current global production of lithium, twenty times the cobalt and forty times the nickel[24] – from mines that do not yet exist, in jurisdictions that are not stable, using energy from the system being replaced.

The wind is free. The turbine is not. Every component is a finite, non-renewable, hydrocarbon-dependent industrial product with a fixed lifespan. And as the next section notes, this is just the first instalment.

THE TREADMILL PROBLEM

A wind turbine lasts twenty to twenty-five years. A solar panel, twenty-five to thirty. A gas plant operates for forty or more. A nuclear plant, sixty to eighty. A renewables-based energy system must therefore be *entirely rebuilt* two or three times within the operational life of the conventional system it replaces – and each rebuild demands the full bill of materials again.

The system does not build once and then produce. It builds, produces, degrades and requires rebuilding – in a cycle that never ends and never gets cheaper in energy terms, because the materials are consumed, not merely borrowed. A turbine blade cannot be recycled into a new turbine blade. The concrete in a foundation is not recoverable. The lithium in a spent battery can be partially reclaimed, but at significant energy cost. This is not maintenance. It is reconstruction.

Now put the three problems together. If a system is on the wrong side of the energy cliff – as the previous section argued wind and solar may well be – it surrenders so much of its gross output to sustaining itself that it must build vastly more capacity to deliver the same net energy as the system it tries to replace. That capacity, spread across wind and solar farms, demands its own vast bill of materials. And that bill must be paid again every generation. The total is the initial buildout, multiplied by the EROEI penalty, multiplied again by the land penalty, then multiplied again by the number of replacement cycles.

The numbers never stop growing, because the cycle never ends.

This is the Renewables Paradox in its full form. A material buildout of unprecedented scale, powered by fossil fuels, dependent on a recursive web of hydrocarbon industry that reaches all the way down, rebuilt every generation, with a net energy return that could easily drive it over the cliff. It is not a transition away from the hydrocarbon system. It is a huge expansion of the hydrocarbon system's workload, undertaken in the name of eliminating it, that has a serious risk of total failure.

But surely there must be an escape – efficiency, hydrogen, green growth? Part V tests the three most popular candidates. Then Part VI follows the consequences into money, and Part VII asks what an honest response would look like.

Part V

The Escape Hatches

In which we test the three most popular arguments that none of the foregoing matters, and find that they don't work either.

By now you may be feeling uncomfortable. The dominant energy supply is depleting, the proposed replacement returns less net energy than the system it replaces, and the physical buildout is beyond anything attempted in human history. Surely there must be an escape.

Enthusiasts of renewable energy systems commonly offer three. The first is efficiency: we will simply use less energy to do the same work. The second is hydrogen: a clean, storable fuel that solves the intermittency problem. The third is decoupling: the economy will grow even as energy consumption falls, because we are shifting to services and information.

Each sounds plausible. None of them works.

THE EFFICIENCY ILLUSION

The most popular response is that we will not need as much energy. We will use what we have more efficiently – better insulation, better engines, better lighting – and do the same work with less fuel. Efficiency will close the gap that physics has opened.

It will not. In 1865, the economist William Stanley Jevons noticed that Britain had dramatically improved the efficiency of its steam engines. Coal was being used more productively than at any point in history. And coal consumption was soaring.

Why? Because more efficient engines made coal *cheaper to use*, which made people use *more* of it – in more places, for more purposes, in industries that had never previously been worth powering. This is the Jevons Paradox, and it has held for 160 years.

Consider lighting. The LED bulb uses roughly 80 per cent less electricity than the incandescent bulb it replaced. If efficiency reduced consumption, global electricity use for lighting should have collapsed. It did not. It rose. Cheap, efficient light made it economical to illuminate car parks, building facades, advertising hoardings and entire city skylines that no one would have lit with incandescent bulbs. The world now has vastly more lit surfaces – and satellite data show that Earth's artificially lit area has been growing at 2.2 per cent per year, with total radiance also increasing. The LED revolution was supposed to reduce energy consumption; instead, cheaper light produced more light.[25]

Or consider cars. American fuel economy standards have roughly doubled the efficiency of new vehicles since 1975. In the same period, total vehicle miles travelled more than doubled. People responded to cheaper motoring not by driving the same distance on less fuel, but by buying heavier vehicles and moving to suburbs that required longer commutes.

The pattern holds at every scale. Economy-wide studies consistently find that more than half of the energy savings predicted by efficiency improvements are consumed by increased activity – a phenomenon economists call the *rebound effect*. The most comprehensive review of the evidence, covering thirty-three studies, found that economy-wide rebound effects typically exceed 50 per cent – with the modelling studies averaging 58 per cent.[26] That means for every ten units of energy that improved efficiency should have saved, only about four were actually saved. And even those savings were overwhelmed by growth – because no country in history has reduced its aggregate energy consumption through efficiency gains while maintaining or growing its economy. Not one. The idea that efficiency can substitute for supply is not supported by a single empirical example anywhere on Earth. It is an article of faith dressed as engineering.

If efficiency cannot close the gap, perhaps a new 'fuel' can. Hydrogen is the favourite candidate.

THE HYDROGEN MIRAGE

Hydrogen is the most abundant element in the universe. It can also be burned, or combined with oxygen in a fuel cell, to release energy without producing CO_2 at the point of use. That is its attraction. It is also the most seductive distraction in energy policy, because it is not a fuel. It is an energy carrier – a way of storing and moving energy that has already been generated elsewhere. Natural hydrogen deposits exist but remain at the earliest stage of exploration. Today, virtually all of the almost one hundred million tonnes produced annually is manufactured – from natural gas and coal – making hydrogen a carrier of fossil energy, not an alternative to it.

And the losses involved in using it are not modest. Start with renewable electricity. Use it to split water in an electrolyser – a device that tears water molecules apart with electricity to release hydrogen gas. You lose energy. Compress the hydrogen for storage: you lose more. Convert it back to electricity through a fuel cell when you need it: you lose more again. Each stage bleeds energy that cannot be recovered. The round-trip efficiency of the full chain – electricity in, hydrogen stored, electricity out – is roughly 30 to 40 per cent.[27] For every three units of renewable electricity you feed into the system, you get roughly one back. The other two are waste heat.

Compare this with a lithium-ion battery, which returns 85–95 per cent of what you put in. Hydrogen does not solve the storage problem. It makes it three times worse.

And the scale required is staggering. To produce that same quantity of one hundred million tonnes from electrolysis using renewable electricity would require roughly 5,000 terawatt-hours of electricity per year – nearly twice the entire annual electricity generation of the European Union.[28] That is before any *new* uses for hydrogen: storage, transport, industrial heat. Just to replace what we already make from fossil fuels would require adding the equivalent of Europe's entire electricity system twice over, and dedicating all of it to hydrogen.

There is also the matter of the infrastructure that does not exist. Hydrogen is the smallest molecule in nature. It leaks through seals, valves and welds that are perfectly tight for natural gas. It diffuses into the crystal structure of steel, embrittling it over time – weakening pipelines, pressure vessels and turbine components in ways that are difficult to detect and expensive to prevent.[29] You cannot simply repurpose the existing gas grid. You would need to largely rebuild it from materials resistant to hydrogen attack, at a cost no government has seriously estimated – because no government has seriously tried. And hydrogen is far more dangerous than natural gas. A pressurised hydrogen leak can ignite spontaneously – no spark, no flame, no external ignition source required.[30]

The hydrogen economy, in short, takes expensive renewable electricity, throws away two-thirds of it, stores the remainder in a gas that destroys its own containers and self-combusts, and calls the result clean energy. It does not solve the intermittency problem. It hides the cost.

Two escape hatches closed. One remains: surely the economy can simply grow its way out?

THE DECOUPLING DELUSION

Every plan for a renewable energy transition rests on an assumption so central that it is rarely stated aloud: that the economy can keep growing while using less energy. This is called decoupling – the idea that GDP and energy consumption, having risen together for the entire history of industrial civilisation, can now be made to go in opposite directions. Green growth. More from less.

The evidence for it is an accounting trick.

When Britain claims that its carbon emissions have fallen 54 per cent since 1990, it is counting only what is burned within its borders. The Office for National Statistics finds that consumption-based emissions, which include the energy embodied in imports, have fallen by just 20 per cent over the same period.[31] And the headline numbers hide a starker fact: UK per-capita electricity generation has fallen by more than a third since its 2003 peak, from roughly 6,700 to 4,200 kilowatt-hours per person.[32] That is not what decoupling looks like. Decoupling means doing more with less energy. Britain has closed its steelworks at Redcar, shut the blast furnaces at Port Talbot, lost its largest aluminium smelter at Lynemouth and has closed the Grangemouth refinery. It is doing less, and importing the rest.

This is not a minor discrepancy. Wiedmann et al., analysing the material footprint of 186 countries, found that the raw materials required to produce the goods that wealthy nations import are, on average, roughly three times the physical weight of those goods when they cross the border – but the extraction, processing and energy consumed to make them never appears in any importing nation's accounts.[33] Their conclusion was blunt: apparent decoupling in advanced economies was smaller than reported, or non-existent.

At the global level, the picture is unambiguous. World primary energy consumption has risen every year for which records exist, barring brief dips during wars and recessions. In 2024, it rose by 2.2 per cent – faster than the average of the previous decade.[34] Energy intensity has fallen, as it should – we get more useful work from each unit of energy than our grandparents did. But what matters to the physics is the total, and the total has never fallen for long – every dip has been erased by the resumption of growth. More efficient use of energy has always led to more consumption, not less. The services economy has not dematerialised economic activity. It has relocated it.

The most comprehensive review of the evidence – 835 empirical studies, analysed by Haberl et al. – found that some countries have reduced their energy use or emissions in absolute terms while continuing to grow, but only slowly. The rate of reduction in those countries falls far short of what climate targets require. The pattern in the data is not that decoupling has been achieved; it is that a small reduction is possible at modest growth rates, but the headline targets cannot be met without curtailing growth itself, or relying on offshoring.[35] When GDP falls, energy use falls. When GDP grows, energy use grows. The relationship has not been broken; it has been hidden behind a shipping container from Shenzhen.

Efficiency increases consumption. Hydrogen destroys more energy than it stores. Decoupling is offshoring with better PR. The escape hatches are closed.

What remains is the question of how to pay for the transition's astronomical costs – and what destroying an energy system does to a financial system.

Part VI

Energy and Your Money

In which we discover that a £10 note is a claim on a future quantity of energy, and that printing more money does not create more energy.

Parts I to V described a physical system: what energy is, what is happening to its supply, why renewables cannot replace it and why the proposed escape hatches do not work. This part follows the consequences into the economy, because the economy is not a separate thing. It is the physical system, counted in money.

Three sections follow. The first argues that money is not wealth but a claim on energy, and that the dominant school of economics has failed to understand this. The second examines the policy that proposes to print money to finance a renewable energy transition, and finds it circular. The third asks what all of this means for your future.

THE ENERGY TOKEN

A ten-pound note does not contain value. It is a *promise* that somewhere in the economy, someone will burn energy to transform raw materials into something you want – a loaf of bread, a hospital visit, an hour of heating. Every transaction in a modern economy is, at bottom, a claim on a future act of energy conversion. Money is the token. Energy is the thing.

This is literal, not figurative. The bread exists because a combine harvester burned diesel, a mill burned gas, an oven used electricity generated by burning gas or coal. Remove the energy and the bread does not exist. The money to buy it becomes a claim on nothing – a piece of paper, or a number on a screen, that refers to something that can no longer be produced.

Neoclassical economics does not see it this way. Its models contain no concept of energy gradient, energy density or thermodynamic constraint. Energy appears as just another commodity – interchangeable with labour, capital or technology. If one input becomes scarce, the model assumes the others will substitute.

This is the assumption on which the entire modern financial system is built: that economic growth can continue regardless of what happens to the energy supply, because human ingenuity will always find a way around physical constraints. Need energy? Just finance it into existence.

This assumption is energy-illiterate, and Parts I to III have shown why. Everything we call civilisation – every city, every factory, every hospital – has been built on a one-time endowment of fossil sunlight that took hundreds of millions of years to accumulate. That energy cannot be created at will, nor can it be substituted. It is the master input – the one without which no other input functions. Labour without energy is manual labour. Capital equipment without energy is scrap metal. Technology without energy is just an idea. Yes, economies have learned to extract more output per unit of energy – but as Part V showed, the apparent 'decoupling' of GDP from energy is largely an accounting illusion, and no economy has ever sustained growth while its energy supply contracted.

The implications for money are immense. Since the economy is, at root, a system for converting energy into goods and services, and money is just a device for keeping score, then the money supply must bear some relationship to the energy supply. When the two move together – as they have throughout the industrial era – money retains its meaning. When they diverge – when the money supply grows while the energy supply contracts – each unit of money claims a smaller share of a shrinking pool of real goods. The name for this is inflation – and when the cause is a structural decline in the energy supply, no monetary policy can resolve it.

FUNNY MONEY

McKinsey, the management consultancy, estimates the capital cost of the renewable energy transition at $275 trillion – roughly $9.2 trillion a year for three decades, or 7.5 per cent of global GDP.[36] And even that, they concede, is likely an underestimate. No government on Earth can raise that sum from taxation, and private capital will not fund a thirty-year, multi-trillion-dollar buildout whose energy economics are, as the previous sections showed, unsettled. So where does the money come from?

The most influential proposal says the question does not matter, because governments that issue their own currency can never run out of money. This is Modern Monetary Theory (MMT), and its core claim is simple: a sovereign government that borrows in its own currency cannot go bankrupt, because it can always create more currency to meet its obligations. Idle resources – unemployed workers, unused factories – can be mobilised simply by spending new money into existence. The only constraint is inflation, and inflation can always be managed by sufficiently wise government policy.

The logic is seductive, and it has become the *de facto* foundation of renewable energy transition financing. Not one country has explained how it intends to pay for it. The unspoken answer is: we will print the money.

The error in MMT is the phrase 'idle resources'. A factory is not idle in the way a banknote in a drawer is idle. A factory is a machine for converting resources into goods. Without energy, the factory is not an idle resource. It is an inert object – a large, expensive building full of metal that does nothing. An unemployed worker is not an idle resource in the relevant sense either. She is a person who can convert food energy into physical or mental labour. But the output of that labour depends entirely on the energy-intensive tools, transport and systems that surround her. In a high-energy economy, one worker with a digger moves a thousand tonnes of earth a day. In a low-energy economy, the same worker with a shovel moves one tonne. The worker has not changed. The energy available to her has. MMT assumes that money can mobilise resources. But resources are mobilised by energy, and money is merely the accounting system that tracks the results.

The circularity is devastating. The renewable energy transition proposes to spend vast amounts of new money – printed, borrowed or conjured from sovereign monetary authorities – to build a new energy system. But the new energy system, as the earlier sections have shown, delivers less net energy than the one it replaces. The money is therefore a claim on an energy supply that is shrinking as a direct consequence of the programme the money is financing. You are borrowing against future economic output to build a system that reduces future economic output. Every pound printed to finance the transition dilutes the energy backing of every pound already in circulation. This is not a theoretical risk. It is an arithmetic certainty. The outcome will be, at best, hyperinflation.

The mechanism is old and well understood: it is what happens when the supply of money grows faster than the supply of real goods and services. What is novel about the renewable transition is the addition of a deliberate constraint on the energy supply that gives the money its meaning. We are not merely printing money. We are printing money while reducing what that money can buy. The arithmetic is unforgiving.

There is a simpler way to say this. You cannot print energy. You cannot legislate energy. You cannot vote for energy. You can only extract it from physical reality according to the laws described in this book, and if your energy policy violates those laws, no quantity of money – however cleverly created – will save you from the consequences.

THE PROMISE

If the previous sections seemed abstract, this one is not. It is about your retirement, and whether you will have one.

Every pension system in the world – state or private, funded or pay-as-you-go – is a claim on future economic output. There is no vault of goods with your name on it. The money in your pension fund is not wealth. It is a claim on goods and services that have not yet been produced – and that can only be produced if the energy to produce them still exists.

When you retire, your income depends on the economy's ability to continue producing real goods and services, and the money to pay for them, for the rest of your life. Pension funds invest in equities, bonds, and property – all of which depend on economic growth for their returns. And it is not just pensions. Every bond, every mortgage, every government debt is a promise that the future economy will produce enough real goods and services to honour it. All debt is a claim on future energy. Since 2008, as Part III noted, the global economy has added roughly $200 trillion in debt and imaginary money. Every dollar of it assumes that the energy to service it will be there. The physics says it will not. The promises remain. The capacity to honour them does not.

The honest name for this is not an energy transition. It is a default on every promise made to every person who expected to retire in comfort.

This is why, in the proposed renewable energy transition,

the financial crisis will arrive before the physical one. Financial systems are forward-looking. Asset prices, currencies and pension valuations will respond to energy contraction before the lights go out. The moment markets understand that the energy supply is in structural decline – and that the policy response is making it worse – the repricing begins. Your money will lose its meaning long before the fuel runs dry.

If that is the diagnosis, the question becomes: what do we do about it?

Part VII

Forging a New Realism

In which we name the energy trap and ask what it would take to escape it.

You now know that the quantity of energy is necessary but not sufficient – that what matters more to our industrial civilisation is its quality: the gradient, the density, the concentration that makes industrial work possible. You know that every successful energy transition in history moved up that quality ladder, and that the proposed transition forces us down it. You know that industrial metabolism requires hydrocarbons as feedstock, not just as fuel. You know that the fossil inheritance is depleting, that the escape hatches do not work and that printing money cannot repeal the laws of thermodynamics.

Every wind farm and solar installation draws down the fossil fuel inheritance: as feedstock to manufacture the devices, and as energy to power the backup that keeps the grid alive when the weather fails. None of it produces a durable replacement. The more aggressively the strategy is pursued, the less surplus remains for a transition that might actually work.

That is the energy trap. And like all traps, it is hidden. The wind farms go up. The subsidies flow. The targets are announced. Everything looks like progress – while underneath, our inheritance burns. And the rough beast that slouches towards us, if the trap is not recognised in time, will not be a managed descent.

Escape from the trap is possible. But three things are required. The first is that there exists exactly one proven technology capable of moving civilisation up the energy ladder: nuclear power. The second is that even with nuclear, some contraction in total energy use is unavoidable – the physics constrains not just what we build but how fast we can build it. The third is that navigating this contraction requires a blueprint built on what the physics permits, not on what the politics would prefer.

And we must also be honest: what follows is without historical precedent. Every previous energy transition replaced a less dense fuel with a more dense one *and* expanded total supply. What the physics now demands is a transition up the quality ladder *while total throughput contracts*. No civilisation has attempted this before. It requires more care, more technical competence and more honesty about physical limits than any energy transition in history. That is the case for taking what follows very seriously indeed.

THE NUCLEAR QUESTION

This book has argued that every successful energy transition in history moved in the same direction: towards steeper gradients, denser fuels and higher power density per unit of land. Wood to coal. Coal to oil. Oil to – what?

The physics answers the question before the politics gets involved. Nuclear fission releases energy by splitting atoms. The energy density of uranium is roughly two million times that of coal and three million times that of wood. A single fuel pellet the size of a fingertip contains as much energy as a tonne of coal, five barrels of oil or 27,000 cubic feet of natural gas. The power density of a nuclear plant – roughly 1,000 watts per square metre of land – matches gas and exceeds every renewable source by two to three orders of magnitude. And it runs continuously, rain or shine, day and night.

By every measure established in Part 1, nuclear is the next rung on the ladder. It is the only energy source available to humanity that is denser, more concentrated and more reliable than fossil fuels. The numbers are not ambiguous.

And the fuel supply, while not infinite, is sufficient. Known uranium reserves, using current light-water reactor technology, could sustain today's nuclear output for a century or more. That figure assumes current consumption – roughly 400 reactors worldwide. A major nuclear expansion would consume reserves faster, and uranium ore grades, like oil field quality, decline over time, raising the energy cost of extraction. These are real constraints, and intellectual honesty requires acknowledging them.

But consider the timescales. The fossil inheritance offers a few decades. Uranium, on current reactor technology and at current scale, offers about a century – though scaled up to replace fossil energy entirely, that horizon shortens, perhaps to decades. What materially extends it are two advanced fuel cycles that are proven physics but not yet commercial: breeder reactors, which exploit the abundant uranium-238 that today's reactors largely ignore; and thorium, three times more plentiful than uranium in the Earth's crust. Both have operated since the mid-twentieth century. Neither has been commercialised at scale. This book will not count unproven capacity as available – but a serious nuclear programme must include commercialising these cycles, because without them, uranium too becomes a bridge fuel rather than a destination. The fossil inheritance offers no such option. Its constraint is geological. Nuclear's is engineering effort. They are different kinds of problem.

So why are we not building the infrastructure?

The honest answer is that nuclear power was defeated not by physics but by politics. Beginning in the 1970s, a sustained and effective campaign – amplified by environmental organisations whose fundraising depended on public fear – succeeded in making nuclear power culturally unacceptable across much of the Western world.[37] The campaign exploited three events: Three Mile Island in 1979, Chernobyl in 1986 and Fukushima in 2011. Each was real. None was what it was claimed to be.

Three Mile Island killed nobody; the partial meltdown was contained by the reactor's safety systems. Chernobyl was a Soviet-era design that no Western country has ever built or would ever license, operated in violation of its own safety procedures. Fukushima was struck by the largest earthquake in Japan's recorded history, followed by a tsunami that killed nearly 20,000 people. The reactor meltdown has since been linked to one worker's death from radiation-induced cancer. The evacuation – ordered in panic and widely regarded as disproportionate – killed over 2,000, mostly elderly people displaced from their homes.

Compare this with the toll of the energy sources that nuclear was prevented from replacing. Fossil fuel combustion kills millions of people a year through air pollution, with coal being the largest single contributor.[38] Hydroelectric dam failures have killed hundreds of thousands. Per unit of energy produced, nuclear power has a death rate orders of magnitude below any fossil fuel and comparable to the safest renewables – even when Chernobyl and Fukushima are included in the count.[39] It is, by any objective measure, among the safest forms of energy generation ever deployed. It was not rejected because it was dangerous. It was rejected because it was feared, and fear is easier to campaign on than arithmetic.

The result is a generation of lost capacity. In the 1970s and 1980s, France built the bulk of its nuclear fleet in fifteen years and today generates roughly 70 per cent of its electricity from nuclear at among the lowest carbon emissions and electricity prices in Europe. Had other Western nations followed that path, the carbon emissions that now dominate the policy debate would already have been reduced by a quantity that dwarfs everything wind and solar have achieved in the decades since.

The cost argument against nuclear is real but misunderstood. Nuclear plants are expensive to build. They are cheap to run. The high construction costs are overwhelmingly a consequence of regulatory complexity that has compounded over decades – not because the reactors are inherently dangerous, but because the regulatory framework became progressively more burdensome in ways that drove up cost and time without commensurate safety benefit. Countries that have maintained standardised, streamlined construction programmes – South Korea, China – build nuclear plants far faster and at far lower cost than the West,[40] at between one sixth and one quarter of the price. The West's inability to do so is a political choice, not a physical constraint.

A new generation of reactor designs – small modular reactors, molten salt reactors, high-temperature gas reactors – promises to reduce construction times, lower capital costs and improve safety margins that are already the best in the energy industry. These are not speculative. Several are in advanced licensing or under construction. They represent the first genuinely new nuclear technology in decades, and they arrive at precisely the moment the physics demands them.

But time matters. A nuclear plant takes seven to fifteen years to build, depending on the regulatory environment. The fossil inheritance, as Part III showed, is no longer expanding – every year of delay narrows the window of bridge fuel available to build the new system. If the remaining hydrocarbon surplus is squandered on energy systems that cannot sustain themselves – the replacement treadmill of Part IV – there may not be enough left to build the one system that can.

Nuclear power is not a complete answer. It generates electricity, and electricity, as Part II showed, cannot replace hydrocarbons as a chemical feedstock. Steel, cement, ammonia and plastics will continue to require fossil inputs for the foreseeable future. What nuclear can do is eliminate the use of fossil fuels for electricity generation – currently the single largest use – freeing the remaining hydrocarbons for the industrial processes that only they can serve. It buys time, and time is the one resource that is running out fastest.

But knowing what to build is not the same as laying foundations. A reactor takes a decade or more to commission. The fossil inheritance is depleting now. Even with a crash programme, there will be a gap – a period where the old system is contracting and the new one is not yet ready. How that gap is managed will determine whether what comes next is a durable civilisation or a chaotic decline. That is the subject of the next section.

MANAGED DESCENT

The previous section identified the gap: a period of a decade or more where the fossil inheritance is contracting and the nuclear replacement is not yet ready to carry the full load. That gap is the consequence of decades of failed planning. The arithmetic merely quantifies what the politics created. And it means that some reduction in the total flow of energy through the economy is unavoidable.

The honest question is not whether living standards will adjust – they will – but whether the adjustment is a managed descent into a durable, lower-throughput civilisation, or an unmanaged collapse into something much worse.

THE INFRASTRUCTURE CANNOT KEEP PACE

The scale of the mismatch between political timetables and physical reality is already visible – not in projections, but in the operational data of real electricity grids. In the United Kingdom, firm generating capacity – the power that is available on demand, regardless of weather – is declining as ageing nuclear and gas plants are retired without replacement. At the same time, policy demands that heating, transport and industry be electrified, which would roughly double peak electricity demand. The gap between what the grid can reliably deliver and what electrification requires is not closing. It is widening.[41]

This is not a peculiarly British problem. Norway – blessed with abundant hydropower, national wealth and strong political commitment – is discovering that even a grid that is almost entirely renewable cannot support full electrification at the pace its government intended: electricity demand is growing six times faster than new generation can be added, and the country is now projected to become a net importer of electricity by around

2030.[42] The Netherlands, having aggressively pursued electrification, is already experiencing grid congestion that prevents expansion. Germany, the most committed renewable builder in Europe, reactivated coal plants in 2022 while simultaneously closing its remaining nuclear reactors,[43] and now acknowledges it needs up to twenty gigawatts of new gas-fired capacity that has yet to be tendered, let alone built.

The pattern is the same everywhere: even wealthy, technically capable nations with favourable geography cannot electrify faster than the physics allows. The political timetable outruns the infrastructure. Targets are set without reference to the generating capacity, grid reinforcement or material throughput needed to meet them. When reality intrudes, the targets are not revised. They are simply missed. The gap is filled with imported fossil fuel – at higher cost, higher emissions and greater geopolitical risk than the domestic production that was shut down.

TWO KINDS OF DECLINE

This exposes a distinction that current policy refuses to make. There are two kinds of energy decline, and they are not the same.

The first is unnecessary decline – decline imposed by ideology. Shutting down domestic oil and gas production while continuing to import the same fuel from elsewhere. Banning technologies that work while subsidising technologies that do not. Setting legally binding targets disconnected from physical reality, then loading the cost of failure onto household energy bills. This decline is a policy choice. It makes the energy position worse without improving anything, because it does not reduce consumption – it merely offshores it. The supply chains that supported domestic production atrophy and cannot quickly be rebuilt; engineering skills and geological knowledge are lost. Meanwhile, imported fuel arrives at higher cost, raising industrial energy prices and eroding the manufacturing base that depends on them.

The second is *necessary decline* – decline imposed by physics. The finite inheritance. The depletion curve. The thermodynamic floor. This decline cannot be wished away. It can only be managed well or badly. A barrel of oil burned today is a barrel unavailable tomorrow, regardless of who burns it or what flag flies over the wellhead.

Current policy imposes unnecessary decline on top of the necessary kind, accelerating the depletion of the inheritance while preventing the investments that might cushion the fall.

The purpose of the blueprint in the next section is to eliminate unnecessary decline and navigate necessary decline. The distinction matters because diagnosis determines treatment. A physician who mistakes a benign cyst for a malignant tumour will prescribe the wrong intervention – and the patient will not survive the error. Current energy policy makes exactly this mistake: it misdiagnoses the condition – and the treatment it prescribes for the wrong disease is destroying the patient's remaining defences against the real one.

WHAT DESERVES INVESTMENT

If the inheritance is finite – and Part III showed that it is – then every barrel of oil, every cubic metre of gas, every tonne of coal represents a choice. The question is not whether to spend it, but what to spend it on. The test is simple: does this investment permanently reduce civilisation's dependence on hydrocarbons, or does it merely shift that dependence onto a parallel system that still requires hydrocarbons to build and maintain?

Nuclear passes it. A reactor consumes hydrocarbon energy during construction – in the steel, the concrete, the enriched fuel. But once operational, it generates electricity for sixty to eighty years with little fossil input. The energy invested is repaid many times over. The demand on the inheritance is a one-time withdrawal, not a recurring levy.

Some renewable technologies qualify – though not the ones that dominate current policy. Geothermal energy, for example, requires a one-time hydrocarbon investment to drill and install. After that, it delivers heat – low-grade heat, the kind needed for space heating and hot water – for decades, with no combustion fuel input, no storage requirement, no weather dependence, and no twenty-five-year replacement cycle.[44] Low-grade heating is the single largest energy demand in many industrial economies. A technology that supplies a substantial fraction of it without hydrocarbons, at the cost of a single drawdown from the inheritance, is precisely the kind of investment that 'protecting the inheritance' should mean.

Better still is the approach that eliminates demand outright. A building constructed to passive-house standards – designed from the outset so that insulation, airtightness and heat recovery do the work – does not merely use energy more efficiently but eliminates the need for most space heating entirely. I built Edinburgh's first certified passive house. It is not theoretical. It is measured, proven and replicable. The modest additional construction cost is repaid within the life of the mortgage; the energy saving continues for the life of the building.

These three strategies – moving up the quality ladder, tapping geological heat and eliminating demand through better buildings – illustrate what elements of a viable response will look like. They are not speculative. Each is proven. Each passes the test: a finite investment from the inheritance that reduces the call on future energy supply once and for all.

Wind and solar fail it. Parts IV and V showed why. They require continuous reinvestment of hydrocarbon energy over their operating life – in manufacturing, in transport, in operation, in maintenance, in the fossil-fuelled backup that must run when the wind does not blow, and in their endless replacements. They do not eliminate demand; they shift it sideways, onto a system that is more materially intensive – far more steel, concrete, copper and rare earths per unit of energy delivered – and just as hydrocarbon-dependent: the fossil backup must still be built and maintained, even when it runs only some of the time. And the whole thing must be rebuilt every generation.

WHAT MANAGED DESCENT LOOKS LIKE

If those are the investments that pass the test, what does the broader picture look like? A managed descent is not a catastrophe. It is a deliberate, planned reduction in total energy throughput, buffered by nuclear baseload and targeted conservation, with the remaining hydrocarbons reserved for the industrial processes that nothing else can support. It is an economy that consumes and wastes less energy.

As the introduction to this part noted, no civilisation has done this before. The transition we propose – to nuclear baseload, conservation and what hydrocarbon use the remaining inheritance can still afford – is harder than anything that has come before. The margin for error is smaller and the consequences of incompetence are worse. But a policy framed around what is measurable today – depletion rates, grid capacity, energy bills – can command public support in a way that one framed around speculative climate predictions never can. People will accept change they can see the reason for.

Living standards adjust. Some of what industrial civilisation currently does will become more expensive or impractical as the cheap energy subsidy of the fossil era withdraws. But the core of what makes life worth living

– warm homes, functioning hospitals, reliable infrastructure, the capacity to feed, educate and care for a population – does not require the energy throughput of the early twenty-first century. It requires *enough* energy, of the right kind, allocated by policy that understands the difference.

That is not a small qualification. Navigating this transition requires something that the current energy debate conspicuously lacks: leadership that understands the physics. People who can read a depletion curve and know what it means. Who can distinguish between a technology that eliminates hydrocarbon demand and one that merely rearranges it. Who understand that energy density sets a floor, and that policy which ignores a floor will collapse.

This book has just spent six parts equipping you to do exactly that. The final section sets out what you should demand – of your government, of your energy companies and of anyone who asks for your vote on the strength of an energy policy. Call it a blueprint. It is built on physics, not on promises.

A BLUEPRINT

Everything in this book reduces to a single observation: energy policy must obey the laws of physics. If it does, options exist. If it does not, no amount of money, legislation or good intention will prevent the consequences. What follows is a seven-point blueprint for an energy policy grounded in physical reality. Each point follows directly from the argument that precedes it. None requires a technological miracle. All require that the people making energy policy understand energy.

The specifics here are written for the United Kingdom, but the physics is universal. Any advanced industrial economy faces the same constraints – depletion, density, thermodynamics – and the same framework applies. The details of domestic geology and regulation will differ; the logic will not.

1. START WITH THE PHYSICS

The thermodynamic floor described in Part 1 is not an arbitrary policy position. It is the ground on which all policy must stand. Energy gradient, energy density and power density set hard limits that no policy can override. Any proposal that contradicts them will not deliver what it promises, regardless of the political will behind it. The first act of an honest energy policy is to require that every energy investment be evaluated against these three measures – and to reject any that risks falling below the threshold at which industrial civilisation can sustain itself.

This implies something uncomfortable: the people setting energy policy must understand the physics, the engineering and the economics. A health secretary is expected to understand public health. A defence secretary is expected to understand military strategy. Yet energy policy – on which everything else depends – is routinely entrusted to politicians who cannot define a kilowatt-hour.

The managed descent described in the previous section is navigable, but not by people who do not understand what they are navigating. In energy policy, competence is a survival requirement.

2. PROTECT AND SPEND THE INHERITANCE WISELY

Part III showed that the fossil endowment is finite and depleting faster than most official projections suggest. This is not a resource to be squandered on energy systems that might return less energy than they consume. Nor is it a resource to be abandoned on principle while the same fuel is imported from elsewhere at greater cost, higher emissions and greater geopolitical risk.

The remaining hydrocarbons are bridge fuel – the energy needed to build whatever comes next. Protecting the inheritance therefore begins with maximising domestic production under sovereign control. The North Sea, where production has declined faster than geology required because of policy choices, should be encouraged to extend its life. Onshore shale gas, of which the UK has substantial potential resources but of which exploration has been banned, should be permitted. Streamlined regulation and a fiscal regime that does not drive investment abroad are essential. The aim is not to maximise consumption, but to ensure that what is consumed during the transition comes from sources we control, at the lowest energy cost, and with the receipts staying in the British economy. How that inheritance is spent is the subject of the next three steps.

3. FAST-TRACK NUCLEAR

The previous section made the physics case. Nuclear fission is the only proven technology that moves up the quality ladder – fusion may follow, but it is not here yet, and policy must be built on what works now. The policy case is equally clear: the regulatory framework that has made nuclear prohibitively slow and expensive in the West is a political artefact, not a physical necessity. Countries that have reformed their regulatory process build reactors on time and on budget.

The United Kingdom should commit to a large-scale nuclear expansion programme, prioritising small modular reactors for speed of deployment and advanced designs for long-term capacity. The barriers that have delayed construction for decades should be dismantled. The distinction is not between 'safety' and 'procedure' – many procedures exist nominally for safety. The right test is whether a given requirement actually reduces real-world risk. Many post-Three Mile Island rules – multi-year licensing iterations for each new reactor design, dose limits set far below natural background radiation, public inquiries that can outlast a project's planning consent – impose enormous cost without commensurate safety benefit. These should go. The genuinely load-bearing safety standards should stay. Delaying now is burning inheritance for nothing.

4. INVEST IN WHAT ELIMINATES DEMAND

The previous section introduced the demand-elimination test: a technology earns investment from the inheritance if it permanently eliminates hydrocarbon consumption, rather than creating another system that still runs on hydrocarbons. At least two further technologies pass this test today.

Geothermal energy and better-insulated buildings both qualify, as the previous section showed. There are many others. Each requires a one-time investment from the inheritance and delivers a lasting reduction in hydrocarbon demand. Public investment should flow to what meets the standard.

5. REDIRECT SUBSIDIES TO WHAT WORKS

Parts IV and V demonstrated that the renewables programme does not pass the demand-elimination test. Its net energy return is too low. Its material demand is too high. Its storage requirements are physically unachievable. And the cost reductions that were promised have not materialised. In the United Kingdom alone, environmental levies on energy bills – subsidies for renewables, capacity payments and related green policy costs – already exceed £16 billion a year, approaching £600 per household, and the UK's independent fiscal watchdog projects them to rise further still, to nearly £20 billion, within five years.[45] The public is paying for an uneconomic system that does not work.

This is not hostility to renewable technology. It is the test, applied consistently. If a technology requires continuous hydrocarbon reinvestment over its operating life, depends on fossil-fuelled backup to maintain grid stability, depends on public subsidy to compensate generators for their uneconomic operations and must be entirely rebuilt every generation, then it fails the test – and subsidising it is not protecting the fossil inheritance but squandering it.

The subsidies, the planning privileges and the levies loaded onto household bills should be redirected to nuclear, geothermal, building insulation and other measures that deliver a lasting return on a finite investment. Public funding for charities and institutions that exist to promote wind and solar deployment should be withdrawn. Existing contracts that commit the public to decades of above-market payments for unreliable energy generation should be renegotiated to reduce the liability.

6. ANCHOR MONEY TO ENERGY

Part VI showed that money is a claim on energy. Since 2008, the global economy has added $200 trillion in debt against an energy supply that is contracting. Print more money than the energy supply can back and the currency loses value. It is that simple.

A government that finances an energy transition that *reduces* the net energy available to its economy is debasing its currency by definition, whether it acknowledges it or not. The consequences are already visible: rising prices, stagnating wages and a standard of living that relentlessly falls even as the economy appears to grow. These are not mysteries. They are what happens when the gap between monetary claims and physical supply grows too wide.

7. TRUST THE PUBLIC WITH THE TRUTH

The United Kingdom audits its money. The Office for Budget Responsibility marks every Budget against reality, and no Chancellor escapes its arithmetic. There is no equivalent for energy – the physical claim behind every pound. There should be: the nation's physical accounts, published every year and independently audited. Net energy delivered. Depletion of domestic reserves. The full system cost of each generating technology – backup, storage and grid included.[46] Dependence on imports. Physical units, not press releases. The three measures of Part I give the auditors everything they need.

This is not a radical demand. It is bookkeeping. A government confident in its energy policy would publish these accounts tomorrow. You have read the book. You know what to look for. Demand them.

Epilogue

When I started the career that led to this book, I had no children. I now have three sons and, so far, two grandchildren. That changes what the word 'inheritance' means to me.

Every barrel of the fossil inheritance wasted on an energy system that cannot sustain itself is a barrel taken from their future. But every barrel now spent wisely – on an energy system that outlasts us – is a gift to their generation, and to the ones after.

No one is coming to fix this. The change will happen because enough people understand the physics and demand that policy obeys it – or it will not happen at all.

The physics is not negotiable. But the response to it is. We can escape the trap.

What's Next?

This book is a starting point, not a conclusion. If you've read this far and you want to do something about it, here's how. Join a growing community of people who are deepening their understanding of energy and thinking hard about solutions. I'm there, I hold regular conversations, and I'd like to meet you.

TheEnergyTrap.org

Give this book to someone. The argument only works if enough people understand it. If this book changed how you think about energy, pass it on.

Glossary

Bridge fuel – The energy needed to sustain civilisation during an energy transition while the new system is being built. Every successful transition in history has required a large surplus of the current source and a denser replacement. The renewable transition fails to meet both conditions.

Capacity factor – The fraction of time an energy source actually produces power at its rated maximum. A gas plant might achieve 85–90 per cent; an onshore wind turbine typically manages 25–35 per cent.

Economic oil – Oil that can be extracted at a price the economy can bear. The distinction between economic and uneconomic oil is the key to understanding 'peak oil'. The planet still contains substantial quantities of hydrocarbons in various forms – bitumen, tight shale, heavy oil – but only the fraction extractable at a tolerable price sustains industrial civilisation. The rest is, for practical purposes, not there at all.

Energy carrier – A means of moving energy from one place to another. Electricity is the most common; hydrogen is another. Unlike a fuel, a carrier cannot be drawn from nature in usable, transportable form – it must be manufactured from a primary source and consumed in real time, always at a conversion loss.

Energy density – The amount of energy stored per unit of mass or volume, typically measured in megajoules per kilogram (MJ/kg). Diesel: around forty-four megajoules per kilogram. Lithium-ion battery: under one megajoule per kilogram.

Energy gradient – The difference in energy concentration between a source and its surroundings (also known in the technical literature as 'exergy'). A steep gradient (e.g. a gas flame at 1,500°C in a 15°C room) means energy can do useful work. A shallow gradient (e.g. a swimming pool at 20°C when the air is the same temperature) means it cannot.

Energy transition – A change in the primary energy source on which a society depends. Historical transitions – wood to coal, coal to oil – each moved to a denser, more concentrated source. The discovery of nuclear fission opened the next step on the same ladder, but it has not been taken at scale. The transition now proposed is unprecedented: it asks us to move in the opposite direction, from concentrated to diffuse.

EROEI – Energy Return on Energy Invested. The ratio of energy delivered by a source to the energy required to build and operate it. A ratio below roughly 7:1 cannot sustain an industrial society.

Fossil sunlight – Hydrocarbons – coal, oil and natural gas – understood as stored solar energy from hundreds of millions of years of photosynthesis, compressed and concentrated by geological time. A one-time endowment, not a renewable flow.

Fuel – A store of energy that exists in nature and can be extracted and used. Coal, oil, natural gas and uranium are fuels. Electricity and hydrogen are not – they are energy carriers that must be manufactured from fuels or other primary sources.

Hydrocarbon – A molecule built from hydrogen and carbon atoms. Coal, oil and natural gas are hydrocarbons. They serve industrial civilisation in two distinct roles: as a fuel (burned to release energy) and as a feedstock (a raw material consumed in chemical processes such as steelmaking, fertiliser production and plastics manufacturing). The second role cannot be replaced by electricity.

Jevons Paradox – The observation, first made by William Stanley Jevons in 1865, that improving the efficiency of fuel use tends to increase total consumption rather than reduce it. Cheaper energy per unit of work makes people do more work. No country has ever reduced its aggregate energy consumption through efficiency gains alone.

Joule – A unit of energy. One joule is roughly the energy needed to lift a small apple one metre. A kilojoule is 1,000 joules; a megajoule is one million joules – roughly the energy in a chocolate bar.

Modern Monetary Theory (MMT) – The claim that a government issuing its own currency can never run out of money, because it can always create more. MMT assumes that 'idle resources' can be mobilised by spending new money into existence. It is the intellectual foundation that has been used to justify net zero financing, but it does not account for the energy required to mobilise those resources.

Neoclassical economics – The dominant school of economics since the late nineteenth century. Its models treat the economy as a system of exchange between rational agents and assume that scarce inputs can always be substituted by other inputs. Energy, when it appears at all, is treated as an ordinary commodity rather than the master input on which all other inputs depend. This book argues that neoclassical economics is energy-illiterate.

Passive house – A building standard that eliminates the need for conventional heating through insulation, airtightness and heat recovery. A passive house typically requires less than fifteen kilowatt-hours per square metre per year for space heating – roughly a tenth of a conventional building. The additional construction cost is repaid within the life of a mortgage; the energy saving continues for the life of the building.

Power – The rate at which energy is converted or transferred, measured in watts. One watt is one joule per second.

Power density – The rate of energy flow per unit of land area, measured in watts per square metre. Gas plants produce around 1,000 watts per square metre, solar farms five to ten watts per square metre, onshore wind farms one to three.

Quality ladder – The pattern, observed in every successful energy transition, of moving to sources with steeper gradients, higher energy density and higher power density. Wood to coal, coal to oil – each step moved up the quality ladder. Nuclear is the next available rung but has not been taken at scale. The proposed renewable transition moves down it.

Watt – A unit of power – one joule per second. A kilowatt is 1,000 watts; a typical UK home draws about 0.3 kilowatts of electricity, or about 1.5 kilowatts including gas heating. A megawatt is one million watts; a gigawatt is one billion; a terawatt is one trillion.

Watt-hour – A unit of energy – one watt of power sustained for one hour, equal to 3,600 joules. A kilowatt-hour is the standard billing unit for electricity (1,000 watt-hours). A typical UK home uses about eight kilowatt-hours of electricity per day. Larger multiples – megawatt-hour, gigawatt-hour, terawatt-hour – scale up by factors of 1,000.

Notes

1 V. Smil, *Power Density: A Key to Understanding Energy Sources and Uses*. The MIT Press, 2015.

2 J. A. Tainter, *The Collapse of Complex Societies*. Cambridge University Press, 1988.

3 World Steel, 'World Steel in Figures 2024'. Accessed: Apr. 03, 2026. [Online]. Available: https://worldsteel.org/data/world-steel-in-figures/world-steel-in-figures-2024/

4 R. M. Andrew, 'Global CO2 emissions from cement production, 1928–2018', *Earth System Science Data*, vol. 11, no. 4, pp. 1675–710, Nov. 2019, doi: 10.5194/essd-11-1675-2019.

5 USGS, 'Mineral Commodity Summaries 2025 – Cement', US Geological Survey, Jan. 2025. Accessed: Apr. 03, 2026. [Online]. Available: https://pubs.usgs.gov/periodicals/mcs2025/mcs2025-cement.pdf

6 FAO, 'Inorganic Fertilizers 2002–2023', Food and Agriculture Organization of the United Nations, 2024. [Online]. Available: https://www.fao.org/statistics/highlights-archive/highlights-detail/inorganic-fertilizers-2002-2023/en

7 Plastics Europe, 'Plastics – the Fast Facts 2024', Plastics Europe, Nov. 2024. [Online]. Available: https://plasticseurope.org/knowledge-hub/plastics-the-fast-facts-2024/

8 V. Smil, *How the World Really Works: The Science Behind How We Got Here and Where We're Going*. Viking, 2022.
9 M. Polidoro, 'Living on air? The crazy ideas and consequences of breatharians', *Skeptical Inquirer*, vol. 44, no. 1, Feb. 2020.
10 IEA, 'World Energy Outlook 2010', International Energy Agency, Nov. 2010. [Online]. Available: https://www.iea.org/reports/world-energy-outlook-2010
11 C. Marsh, 'The remarkable expansion of global central bank balance sheets during the pandemic', *Money: Inside and Out*. Accessed: Apr. 03, 2026. [Online]. Available: https://moneyinsideout.substack.com/p/the-remarkable-expansion-of-global
12 IIF, 'Global Debt Monitor', Institute of International Finance, Feb. 2026. [Online]. Available: https://www.iif.com/Publications/publications-filter/c/Global%20Debt%20Monitor
13 EIA, 'U.S. Field Production of Crude Oil (Thousand Barrels per Day)', US Energy Information Administration. Accessed: Apr. 03, 2026. [Online]. Available: https://www.eia.gov/dnav/pet/hist/leafhandler.ashx?n=pet&s=mcrfpus2&f=m
14 IEA, 'Global Energy Review 2025', International Energy Agency, Mar. 2025. [Online]. Available: https://www.iea.org/reports/global-energy-review-2025
15 C. Campbell and J. Laherrère, 'The end of cheap oil', *Scientific American*, vol. 278, no. 3, pp. 78–83, Mar. 01, 1998.
16 Rystad, 'World's proven oil reserves equal 14 years of production'. [Online]. Available: https://dieselnet.com/news/2025/08rystad.php
17 V. Smil, *Energy Transitions: History, Requirements, Prospects*. Prager, 2010.

18 C. A. S. Hall, J. G. Lambert, and S. B. Balogh, 'EROI of different fuels and the implications for society', *Energy Policy*, vol. 64, pp. 141–52, Jan. 2014, doi: 10.1016/j.enpol.2013.05.049.
19 C. Llewellyn Smith, 'Large-Scale Electricity Storage', The Royal Society, Sep. 2023. Accessed: Sep. 04, 2024. [Online]. Available: https://royalsociety.org/news-resources/projects/low-carbon-energy-programme/large-scale-electricity-storage/
20 D. Weißbach, G. Ruprecht, A. Huke, K. Czerski, S. Gottlieb, and A. Hussein, 'Energy intensities, and energy payback times of electricity generating power plants', *Energy*, vol. 52, pp. 210–21, Apr. 2013, doi: 10.1016/j.energy.2013.01.029.
21 C. J. Barnhart, M. Dale, A. R. Brandt, and S. M. Benson, 'The energetic implications of curtailing versus storing solar- and wind-generated electricity', *Energy and Environmental Science*, vol. 6, no. 10, pp. 2804–10, Sep. 2013, doi: 10.1039/C3EE41973H.
22 G. Palmer, 'A framework for incorporating EROI into electrical storage', *Biophysical Economics and Resource Quality*, vol. 2, no. 2, p. 6, Apr. 2017, doi: 10.1007/s41247-017-0022-3.
23 IEA, 'The Role of Critical Minerals in Clean Energy Transitions', International Energy Agency, May 2021.
24 ibid.
25 C. C. M. Kyba et al., 'Artificially lit surface of Earth at night increasing in radiance and extent', *Science Advances*, vol. 3, no. 11, p. e1701528, Nov. 2017, doi: 10.1126/sciadv.1701528.
26 P. E. Brockway, S. Sorrell, G. Semieniuk, M. K. Heun, and V. Court, 'Energy efficiency and economy-wide rebound effects:

A review of the evidence and its implications', *Renewable and Sustainable Energy Reviews*, vol. 141, p. 110781, May 2021, doi: 10.1016/j.rser.2021.110781.

27 K. Mongird, V. Viswanathan, J. Alam, C. Vartanian, V. Sprenkle, and R. Baxter, '2020 Grid Energy Storage Technology Cost and Performance Assessment', US Department of Energy, Technical DOE/PA-0204, 2020.

28 IEA, 'The Future of Hydrogen – Analysis', International Energy Agency Jun. 2019. Accessed: Apr. 04, 2026. [Online]. Available: https://www.iea.org/reports/the-future-of-hydrogen

29 A. Laureys, R. Depraetere, M. Cauwels, T. Depover, S. Hertelé, and K. Verbeken, 'Use of existing steel pipeline infrastructure for gaseous hydrogen storage and transport: A review of factors affecting hydrogen induced degradation', *Journal of Natural Gas Science and Engineering*, vol. 101, p. 104534, May 2022, doi: 10.1016/j.jngse.2022.104534.

30 G. R. Astbury, 'A review of the properties and hazards of some alternative fuels', *Process Safety and Environmental Protection*, vol. 86, no. 6, pp. 397–414, Nov. 2008, doi: 10.1016/j.psep.2008.05.001.

31 A. Syed, 'The Decoupling of Economic Growth from Carbon Emissions: UK Evidence', Office for National Statistics, Oct. 2019. ONS, 'Greenhouse gas emissions and trade, UK: 2024', https://www.ons.gov.uk/economy/environmentalaccounts/articles/greenhousegasemissionsandtradeuk/2024

32 Energy Institute, 'Statistical Review of World Energy 2025', Jun. 2025. [Online]. Available: https://www.energyinst.org/statistical-review

33 T. O. Wiedmann et al., 'The material footprint of nations', *Proceedings of the National Academy of Sciences*, vol. 112, no. 20, pp. 6271–6, May 2015, doi: 10.1073/pnas.1220362110. See also D. Helm, *Net Zero: How We Stop Causing Climate Change* (William Collins, 2020), which argues for measuring carbon on a consumption rather than a territorial basis.

34 See note 14 above.

35 H. Haberl et al., 'A systematic review of the evidence on decoupling of GDP, resource use and GHG emissions, part II: synthesizing the insights', *Environmental Research Letters*, vol. 15, no. 6, p. 065003, Jun. 2020, doi: 10.1088/1748-9326/ab842a.

36 McKinsey, 'A Net-Zero Economy: The Impact of Decarbonization', McKinsey Sustainability, Jan. 2022. Accessed: Jan. 25, 2023. [Online]. Available: https://www.mckinsey.com/capabilities/sustainability/our-insights/the-economic-transformation-what-would-change-in-the-net-zero-transition

37 G. Cravens, *Power to Save the World: The Truth About Nuclear Energy*. Knopf, 2008.

38 K. Vohra, A. Vodonos, J. Schwartz, E. A. Marais, M. P. Sulprizio, and L. J. Mickley, 'Global mortality from outdoor fine particle pollution generated by fossil fuel combustion: Results from GEOS-Chem', *Environmental Research*, vol. 195, p. 110754, Apr. 2021, doi: 10.1016/j.envres.2021.110754.

39 H. Ritchie, 'What are the safest and cleanest sources of energy?', *Our World in Data*, Feb. 2020. [Online]. Available: https://ourworldindata.org/safest-sources-of-energy

40 IEA, 'Projected Costs of Generating Electricity 2020', International Energy Agency, Dec. 2020 [Online]. Available:

https://www.iea.org/reports/projected-costs-of-generating-electricity-2020; J. R. Lovering, A. Yip, and T. Nordhaus, 'Historical construction costs of global nuclear power reactors', *Energy Policy*, vol. 91, pp. 371–82, Apr. 2016, doi: 10.1016/j.enpol.2016.01.011.

41 K. Porter, 'Electrification – can the grid cope?', *Watt-Logic*, Jan. 2026. [Online]. Available: https://watt-logic.com/2026/01/12/electrification-can-the-grid-cope/

42 DNV, 'Norway's green energy industry sector stalling – industry the loser in tight power market', press release on Energy Transition Outlook Norway 2024, Dec. 03, 2025. Accessed: May 05, 2026. [Online]. Available: https://www.dnv.com/news/2025/norways-green-energy-industry-sector-stalling-2/.

43 WNA, 'Nuclear Power in Germany', World Nuclear Association, Mar. 2026. [Online]. Available: https://world-nuclear.org/information-library/country-profiles/countries-g-n/germany.

44 Arup, 'UK Geothermal Review and Cost Estimations', Department for Energy Security and Net Zero, May 2025.

45 OBR, 'Economic and Fiscal Outlook – March 2025', Office for Budget Responsibility, Mar. 2025. [Online]. Available: https://obr.uk/efo/economic-and-fiscal-outlook-march-2025/

46 The principle that the whole-system cost of generation – backup, storage and network included – should be made transparent and charged to those who impose it is central to Dieter Helm; see his *Cost of Energy Review* (Department for Business, Energy & Industrial Strategy, 2017) and *Net Zero* (2020).

Further Reading

The references in the notes of this book document specific claims. This section is different: it is a short guide, for the reader who wants to go further, to the books and reports that shaped the argument. They are grouped loosely by the themes of the book they illuminate. All are accessible to the general reader unless noted otherwise.

ENERGY AND CIVILISATION

Vaclav Smil, *Energy and Civilization: A History* (MIT Press, 2017). The definitive account of how energy sources have shaped human societies from biomass to fossil fuels. Smil's quantitative rigour is unmatched; no serious discussion of energy policy should begin without it.

David J. C. MacKay, *Sustainable Energy – Without the Hot Air* (UIT Cambridge, 2009). A Cambridge physicist asks a deceptively simple question – can renewable energy power Britain? – and does the arithmetic honestly. MacKay's optimism about engineering solutions makes his conclusion all the more striking: even on generous assumptions, the numbers are not close. Fifteen years on, the technologies have improved but the underlying arithmetic – areal power density, storage at scale, the gap between supply and demand – has not. The case has, if anything, hardened. Essential reading for anyone who wants to check the numbers in Parts IV and V of this book. Available free at withouthotair.com.

Vaclav Smil, *Power Density: A Key to Understanding Energy Sources and Uses* (MIT Press, 2015). The concept that organises much of the argument in Parts I and IV. Smil demonstrates why power density – the rate of energy flow per unit of land – is the decisive constraint on infrastructure, land use and the viability of energy transitions.

DEPLETION AND LIMITS

Kenneth S. Deffeyes, *Hubbert's Peak: The Impending World Oil Shortage* (Princeton University Press, 2001). Written by a Princeton geologist who worked alongside M. King Hubbert, one of the godfathers of the peak oil theory. A clear, short account of why oil production must peak and what the geology actually says. His timing was early; his geology was right.

William R. Catton Jr, *Overshoot: The Ecological Basis of Revolutionary Change* (University of Illinois Press, 1980).

The foundational argument that industrial civilisation has exceeded the long-term carrying capacity of its environment by drawing down a one-time inheritance of fossil energy. Catton's framing – that we are *Homo colossus*, a species temporarily inflated by a non-renewable subsidy – remains the clearest statement of the predicament this book describes.

NET ENERGY AND THE RENEWABLE QUESTION

Charles A. S. Hall and Kent Klitgaard, *Energy and the Wealth of Nations: An Introduction to Biophysical Economics* (2nd edn, Springer, 2018). The textbook on energy return on energy invested and its economic consequences. More academic than most entries here, but the framework is indispensable for understanding why net energy – not gross output – determines what a civilisation can sustain. Provides the analytical foundation for Parts IV, V and VI.

COMPLEXITY AND COLLAPSE

Joseph A. Tainter, *The Collapse of Complex Societies* (Cambridge University Press, 1988). The most rigorous analysis of why complex societies simplify. Tainter's thesis – that collapse is a rational response to diminishing returns on complexity – provides the theoretical foundation for understanding why managed descent, not technological rescue, is the appropriate response to energy contraction.

NUCLEAR ENERGY

Gwyneth Cravens, *Power to Save the World: The Truth about Nuclear Energy* (Alfred A. Knopf, 2007). A former nuclear sceptic visits every stage of the fuel cycle and reports what she finds. Accessible, honest and considerably more persuasive than any advocacy document. A useful corrective for readers whose instincts on nuclear power were shaped by decades of campaigning rather than evidence.

WHAT COMES NEXT

John Michael Greer, *The Long Descent: A User's Guide to the End of the Industrial Age* (New Society Publishers, 2008). The most sustained argument that industrial decline will be a long staircase rather than a single cliff. Greer draws on an unusually wide range of historical and philosophical traditions; his analysis of how civilisations contract in stages is sharp regardless of where one stands on his broader worldview.

James Howard Kunstler, *The Long Emergency: Surviving the End of Oil, Climate Change, and Other Converging Catastrophes of the Twenty-First Century* (Atlantic Monthly Press, 2005). The subtitle reflects the assumptions of the time; Kunstler's lasting contribution is his analysis of oil depletion, infrastructure decay and the fragility of systems that have no fallback. He writes with a journalist's eye for the telling detail and a willingness to follow the logic to conclusions that polite opinion prefers to avoid.

ENERGY AND MONEY

Frederick Soddy, *Wealth, Virtual Wealth and Debt: The Solution of the Economic Paradox* (Allen & Unwin, 1926). A Nobel laureate in chemistry turns his attention to economics and discovers that orthodox theory ignores the laws of thermodynamics. Soddy demonstrates that real wealth is physical and subject to entropy, while financial debt compounds without limit – a divergence that must eventually be resolved by default, inflation or deliberate policy. Written a century ago, it remains the most fundamental critique of the assumption that money can grow independently of physical resources. Demanding but foundational; Part VI of this book is deeply indebted to it.